NATIONAL LITERACY TESTS Y7

A+ National Practice Tests
graduated difficulty with answers

Lynne Marsh
Wendy Bodey
Virginia Ayliffe
Julie Mitchell

A+ National

NELSON
A Cengage Company

National Literacy Tests Year 7
1st Edition
Lynne Marsh
Wendy Bodey
Virginia Ayliffe
Julie Mitchell

Publishing editors: Jana Raus, Jane Moylan
Project manager: Mia Yardley
Senior designer: Ami Sharpe
Text designer: Ami Sharpe
Cover designer: Ami Sharpe
Cover image: iStockphoto
Photo researcher: Libby Henry
Production controller: Alex Ross & Damian Almeida
Reprint: Magda Koralewska
Reprint: Alice Kane
Reprint: Katie McCappin

Any URLs contained in this publication were checked for currency during the production process. Note, however, that the publisher cannot vouch for the ongoing currency of URLs.

Acknowledgements
VCE ® is a registered trademark of the VCAA. The VCAA does not endorse or make any warranties regarding this Cengage product. Current and past VCE Study Designs, VCE exams and related content can be accessed directly at www.vcaa.vic.edu.au

We would like to thank the following for permission to reproduce copyright material:

AMB Productions: p. 20 (images); © Commonwealth of Australia, reproduced with permission: p. 20 (text); Corbis Australia: p. 48; © The Duke of Edinburgh's Awards of Australia: 85 (logo & text); © State of Victoria (Department of Education and Early Childhood Development): p. 82; © 2009 Emerald Tourist Railway Board: pp. 87 – 88; Sofia Fitri: pp. 18, 19; iStockphoto: pp. 67 Jan Rysavy, 51 Jane Norton, 85 (image) Paul Morton; Jupiterimages Corporation: pp. 31, 46 (top); © Lord Howe Island Tourism Association: p. 50; © 2007 Metro Magazine: p. 49; Cheryl Nicholson: p. 101; Reproduced with permission by Penguin Group (Australia): pp. 53, 86 (text); Shutterstock: pp. 22 Ian Scott , 46 (bottom) Jacek Chabraszewski, 81 (image) Maksym Gorpenyuk, 45 michaeljung, 86 (image) urosr, 79 Yuri Arcurs; The Science Museum (www.sciencemuseum.org.uk): p. 47.

P 80: 'Bora Ring' by Judith Wright published in Hamilton, Elaine and Farr, Robin. Poetry Unlocked, Farr Books, Queensland. 2008, p 94.

P 81 (text): Ancient Egypt http://en.wikipedia.org/wiki/Ancient_Egypt). Available under the GNU Free Documentation License (http://en.wikipedia.org/wiki/Wikipedia:Text_of_the_GNU_Free_Documentation_License)

Every effort has been made to trace and acknowledge copyright. However, if any infringement has occurred the publishers tender their apologies and invite the copyright holders to contact them.

For product information and technology assistance,
in Australia call **1300 790 853**;
in New Zealand call **0800 449 725**

For permission to use material from this text or product, please email **aust.permissions@cengage.com**

ISBN 978 0 17 046219 8

Cengage Learning Australia
Level 7, 80 Dorcas Street
South Melbourne, Victoria Australia 3205

Cengage Learning New Zealand
Unit 4B Rosedale Office Park
331 Rosedale Road, Albany, North Shore 0632, NZ

For learning solutions, visit **cengage.com.au**

Printed in China by 1010 Printing International Limited.
2 3 4 5 6 7 25 24 23 22

NATIONAL LITERACY TESTS Y7

Detailed information

Introduction

Literacy and numeracy are the fundamental building blocks of learning in any subject. Knowing what you can do and where you need to improve is vital for all Australian students, teachers and parents. The NAPLAN* (National Assessment Program – Literacy and Numeracy) tests in literacy and numeracy help governments find out how Australian students are progressing and help to identify what you know and what you don't know. The results of the tests also help you and your teachers plan what you need to learn next.

Tests can sometimes be a little daunting; however, there are practical steps you can take to ensure you will successfully sit a test. You will be best prepared for any test, when you understand what is being tested, how you will be tested, and when you are mentally and physically prepared for the test. This book provides practical advice and strategies to ensure that you are test ready and contains practice tests so you can see what to expect in the tests.

This book and the accompanying NelsonNet website provide:

- ✔ **Test tips:** advice on how to successfully sit tests, information about what is tested and how it is tested, hints on responding to different question types and how to act on your results.
- ✔ **Key skills:** summary information about each of the three Literacy tests: Reading, Language Conventions and Writing. This includes important information about what each of the tests cover, including the skills, knowledge and understanding you may be required to demonstrate. You will also find examples of actual student responses demonstrating different levels of understanding.
- ✔ **Practice question sets:** includes practice question sets for each of Language Conventions (spelling, grammar and punctuation), Reading and Writing. These tests are designed so that you can build your skills and confidence during the first part of the year leading up to the test.
- ✔ **Two full-length practice tests:** two full-length detachable tests for each of Language Conventions (spelling, grammar and punctuation), Reading and Writing. One paper is for you to use as a practice test, and the other is for you to hand in to your teacher. These tests follow the format of the NAPLAN* tests and are the level of difficulty you can expect to find when you sit the actual tests. These are ideal for practice during the month leading up to the test.
- ✔ **Answers:** for easy reference, answers to all workbook questions can be found on NelsonNet, https://www.nelsonnet.com.au/free-resources.
- ✔ **Useful icons:** icons are used throughout the resources to help you quickly find other important information in the resources and support your understanding.

HOT TIP	**Hot tips:** keep one step ahead by heeding these hot tips. The tips often relate to things many students forget to do or can do better with a little planning.
world wideweb	**World Wide Web link:** provides a useful link to further information available on the Internet.

* **This book is not an officially endorsed publication of the NAPLAN project and is produced by Cengage Learning independently of Australian governments.**

About the Authors

Lynne Marsh

Lynne Marsh is our lead author for both of the Y7 and Y9 books, and joins us with a wealth of teaching and assessment experience from NSW. She has been a co-author of *The Text Book* series and several resources for The English Teachers Association and the NSW Department of Education and Training, and is currently teaching in NSW.

Wendy Bodey

Wendy Bodey draws on diverse experience gained working at Curriculum Corporation, the Australian Council for Educational Research and as a teacher. Wendy has authored a variety of assessment materials and resources and has played key roles in national and jurisdiction assessment programs including the inaugural National Assessment Program Literacy and Numeracy (NAPLAN) tests.

Virginia Ayliffe

Virginia Ayliffe is a co-author on the renowned *Shakespeare Unplugged* series, which last year won the APA Educational Excellence Award for Best Secondary Series. She currently teaches at Somerville House in Queensland.

Julie Mitchell

Julie Mitchell has authored and compiled several English texts, and has many years' experience teaching secondary English in Victoria. She has also written curriculum resources in the field of Values Education with Curriculum Corporation. She is now involved in teaching preservice teachers at the University of Melbourne.

Test Tips

All tests are designed to provide useful information to help you and your teacher plan what you need to learn next. The NAPLAN* tests are no different. All you need to do is be as prepared as you can and answer the questions to the best of your ability. Follow the practical advice provided in the rich resources throughout this book and you will be prepared to successfully sit most tests at the secondary level.

1 HOT TIP

Know your tests

It is good to know as much as you can about the tests so that you have an understanding of what is being tested and how it will be tested. There are three Literacy tests, one for each of Reading, Language Conventions and Writing. Language Conventions incorporates Spelling, Grammar and Punctuation. Understanding more about the types of skills assessed in each test is useful information for you as a test taker. Use the resources provided to find out more about the composition of the tests including the content coverage, the aspects included and the marking criteria used.

2 HOT TIP

Know the type of answer required

The Writing test requires you to write on a set topic. There is quite a difference between an average and an excellent response. Follow the advice provided in these resources, particularly the advice provided in the Writing Key Skills section. This section includes annotated examples so that you can see how the marking criteria are applied to actual student responses. The Reading and Language Conventions test questions cover a range of aspects and difficulty levels and generally the questions become more difficult as you proceed through each test. The Language Conventions and Reading Key Skills sections include important rules, conventions and other information that you will find useful during the tests.

Many of the questions in the Language Conventions test and the majority of those in the Reading test are set out in multiple-choice format. The correct answer is provided along with some other, often attractive, options for you to choose from. You need to identify the correct response.

The Reading test questions relate to texts provided in the reading stimulus booklet. A small number of these questions require you to write your answer. These questions usually require you to provide an explanation or evidence from the text to support your answer to the question. Your answer will be marked based on the understanding you demonstrate so it is important your answer is clear and unambiguous.

3 HOT TIP

Tackling multiple-choice questions

Multiple-choice questions provide the correct answer along with some other options for you to choose from. It is important that you try to work out the correct answer before studying the answer options. In this way you will not be distracted by other attractive, but incorrect, options.

Consider the following question and then read the recommended approach that you can apply to any question.

Which of the following correctly completes the sentence?

The boy ☐ loudly.

☐ sing ☐ sang ☐ sung ☐ singed

First, try to work out the correct answer without closely considering the available options. All options relate to singing so try saying the sentence to yourself to help you decide on the best word to complete the sentence. You might think, 'The boy was singing loudly' or 'The boy sings loudly', or, 'The boy sang loudly'. Next read the options provided. The option 'sang' is available so this is likely to be the correct answer. When you have confirmed all the other options are incorrect you can then make your final selection.

If you are unsure of the answer to a question try eliminating any obviously incorrect options. In the question above the option 'singed' can be easily eliminated as there is no such word. The option 'sing' can be eliminated as its singular form is inappropriate in this sentence. The tense of 'sung' is also not correct; leaving 'sang' the remaining option. Try out the remaining option – 'The boy sang loudly'. This check confirms you have the correct answer. If your check is not conclusive, consider coming back to the question after you have completed the rest of the test.

4 HOT TIP

Make sure your answers show your understanding

When you answer a question, make sure your answer is

* **This book is not an officially endorsed publication of the NAPLAN project and is produced by Cengage Learning independently of Australian governments.**

clear and unambiguous. Consider the following Reading question and the selection of possible responses. This is an approach you can apply to any question that requires you to demonstrate an understanding of the text.

First you will be directed to the passage you are required to answer questions about, for example:

Read 'Frogs Return' and then answer the following question.

For this purpose we will imagine you have read the text. The text describes the return of frogs to the Wilson Creek area after an absence of 15 years. Scientists regard the presence of frogs in an area as one of the best indicators of a healthy environment. Scientists credit the improvement at Wilson Creek to work by local community groups to restore the area and are pleased the creek can again sustain frog life.

You will then be asked questions about the text, for example:

Why were the scientists excited about the presence of frogs at the creek?

Student One's answer: They are important.

This response is too vague and does not clearly convey why the scientists would be excited about the frogs' presence. An answer of this quality would be marked incorrect.

Student Two's answer: Not seen there for years and a sign of good environment.

This answer includes an explanation for the scientists' excitement – the return of frogs and this being evidence of an improvement in the environment. As this answer demonstrates an accurate understanding of the text it is credited as correct.

Remember you will not be penalised for incomplete sentences, or indeed, incorrect spelling, grammar or punctuation. You should, however, ensure your answer is clear and legible so that the marker can understand your response and credit the understanding you have shown.

5 HOT TIP

Watch the time but don't hurry

The time available to complete each of the tests is adequate for most students to comfortably complete the test. So be conscious of the time as you work through the test but don't rush your answers.

You have 40 minutes to complete the Writing task including an allowance of 5 minutes for planning at the start and 5 minutes of editing at the end of the session. It is important to plan what you are about to write if you are then to make the most of your 30 minutes' writing time. It is also important you make use of the last 5 minutes to reread your work and make any necessary corrections.

In the Reading test you have just over 1 minute per question and in the Language Conventions test you have about 1 minute per question. Start by working through the questions you are most confident to answer and do quick checks as you go. Use the remaining test time to tackle questions you are less sure of. Do not spend too much time on any one question until you have completed all other questions in the test.

6 HOT TIP

Prepare well

Some students feel nervous when it comes to test time. The best way to manage this is by ensuring you are well prepared before test day, and then you will have no need for any concern. Find out as much as you can about each test well before you sit them so that you have an understanding of what is being tested, and how it will be tested. Read through the various Key Skills sections and the glossary to revise important facts, rules and other information that you will find useful during the test and in building your knowledge.

Work through the Spelling, Punctuation, Grammar, Reading and Writing practice sets provided to give you a better sense of what the test will be like and to assist you to identify any areas that you may need to revise. Use these sessions to monitor how well you use your time.

Then complete the two full-length detachable tests incorporating Language Conventions, Reading and Writing. The first is designed for you to use as a practice test, the other to hand in to your teacher. Discuss your test results with your teacher and seek their feedback on how well you performed and any areas you might improve on.

7 HOT TIP

Be calm and don't panic

On test day make sure you have had a good night's sleep and that you have eaten breakfast so that you are physically prepared to do the test. Be confident in your preparation and you will be ready to tackle the test. Use the pretest and test-day checklists provided to assist your preparation.

It is important that you answer each test question to the best of your ability as this will provide the best guide for your future learning requirements. Tests help to identify your strengths and any potential areas for further development that you may have. It is important that you discuss your results with your teacher to ensure your learning program and goals are the most appropriate for you.

8 HOT TIP

Enjoy adding to your skills

There is a well-known saying that practice makes perfect and nowhere is this more evident than in the use of language. By including activities such as reading, completing word puzzles and learning new words in your everyday life you will build your literacy skills without even thinking about it. Reading is an excellent and enjoyable way to build your vocabulary, to see the way other writers craft their work and to learn. Reading books, newspapers, magazines and website information is not only educational but can also be very relaxing. It is useful to expand the range of material you read. If you currently read mostly informational texts then it's time to try some different genres such as science fiction, historical fiction – see if your library has a list of recommended books to guide you in a new direction. Have a go at different types of word puzzles and games such as: crosswords, anagrams and scrabble. Learn a new word every day; many newspapers run a word of the day column. Find out what the word means and how to use it. Include new words in your writing where appropriate so that your writing reflects a broad and interesting vocabulary. Focus on providing alternative choices for words that tend to be repeated a lot like 'then', 'after' and 'later'.

Play 'spot the mistake' with your family when you're out: see who is first to find a spelling error. It might be on a shop's sign, in a cafe menu or on a billboard.

Read out loud to your family or to a younger sibling to practise your fluency and check you understand what you are reading. Reading short excerpts from the newspaper can be a good discussion starter.

So incorporate informal activities such as these in your day and then sit back and relax while you learn.

▶ 9 HOT TIP

Know what you're talking about

Do your eyes glaze over when you hear words such as 'alliteration' or 'metaphor'? Then you need to find out what these words mean, as chances are you are not currently able to capitalise on the use of these in your writing. The Glossary is a good starting point as this lists important words you are likely to need to know. Take the opportunity to learn any words that you don't know and to revise those words that you have met previously. The Key Skills sections will exemplify some of the techniques or types of language that it is useful for you to know about. This will give you a context to understand the use and effect of various writing techniques. Remember writers craft their work – making deliberate word and structure choices to give an intended effect. For you to have such control in your own writing you need to develop these skills and understand the way you can create certain effects. It is important to experiment with new techniques well before the test so that you are confident in your use of the technique before you might choose to use it in a test situation. It is not a good idea to try out a new technique in the test as it sometimes takes a few attempts to create the impact you are hoping to achieve.

▶ 10 HOT TIP

Adjusting to the time available

As discussed in Tip 5 the time provided for each test is adequate for most students. There are, however, things you can do to ensure you make the most of the time available.

Have a watch with you so that you are not dependent on time information provided at the venue. Forty minutes to complete a writing task might be quite a lot less time than you would normally spend so make sure you have completed the practice tests and have a good sense of what you can achieve in the time allowed. The first 5 minutes is for planning your response. Planning what you can realistically achieve in the time available is the most valuable use of this time. You are limited in the number of pages for your response; this reinforces a need to keep your writing contained. Otherwise you might find you begin to ramble and run out of time or space.

It is important that you use the final 5 minutes to review and if necessary revise your work so that it reads fluently and captures an audience's interest. You may not have time to finely polish your text but it does need to be coherent and complete.

Similarly in the Reading and Language Conventions tests you need to be conscious of your time use.

You have about a minute to answer each Reading question and this includes reading the text. Underline important information in each text as you read so that you have handy reference points to refer to when you come to answer the questions. If you are having trouble answering a question move on and come back to it later – it's a good idea to circle the question number to tag where you need to return to. If you are struggling to understand a particular text, move on and come back to it after you have completed as much of test as you can.

You have just under a minute to answer each Language Conventions question; this sometimes requires reading a short text before the questions. Sometimes this requires reading answer options that are quite long. Usually in these cases each sentence is very similar and you might, for example, be asked to identify the correct punctuation. Once you know what the sentence is you simply need to find the one that is correctly punctuated. Know what you are looking for: in this example it could be the correct location of quotation marks. Focus on finding what you are looking for quickly. Once the correct answer is located it will not take long to check that the other options are incorrect so that you can confirm your final option choice.

▶ 11 HOT TIP

Answer position is important

Where you write your answer can impact on your results. Many of the questions you answer will be

9780170462198

machine scored after the test so it is important that you write your answer in the box provided or on the lines provided and that you do not write in page borders or on other parts of the page. It is important in your Writing response that you stay within the set page limits you are given. Any attached pages are unlikely to be considered when marking your response.

When responding to multiple-choice questions make sure you only colour in the number of boxes required and that you colour them in completely. Don't just draw a line through the box as the machine may not detect this. Most questions have one answer; however, there are some questions where you are asked to colour in two boxes. An instruction icon will tell you how many boxes to colour in. If you want to change your answer make sure you rub out the other answer completely so that the machine does not accidentally record more answers than you meant to provide.

Make sure your written answers are legible. If it is difficult to read your answer you may not get credited for a correct response.

12 HOT TIP

Keep up to date with information from your test authority

It is important to keep up with any information about the test. Your test authority will provide regular updates. Contact details for all Australian Test Administration Authorities for the NAPLAN* tests can be found at:

www.naplan.edu.au/test_administration_authorities.html

Other general information about the tests and specific information such as the genre for this year's writing test can be found at:

www.naplan.edu.au/

* This book is not an officially endorsed publication of the NAPLAN project and is produced by Cengage Learning independently of Australian governments.

Pre-test checklist

Use this checklist to ensure you are prepared to successfully sit the tests.

	Activity	Main Resource
❐	Know the important words and terms used in literacy	Glossary
❐	Be familiar with the keys skills for each test – Reading, Writing, Language Conventions (Spelling, Grammar and Punctuation)	Key Skills
❐	Be familiar with English conventions	Key Skills
❐	Build your skills and knowledge informally too	Read widely Tackle puzzles and games Learn word definitions
❐	Understand the Writing marking criteria and how they are applied	Writing marking criteria Writing Key Skills
❐	Complete the practice question sets to get a sense of the test content and questions types	Complete practice set questions for each test
❐	Check your answers against the solutions to evaluate your strengths and any areas you need to revise	Practice question sets solutions for each test
❐	Get advice on tackling the tests	Test tips
❐	Complete Practice test 1, the same length and level of difficulty you can expect in the test	Practice test 1 for each test
❐	Check your answers against the solutions to evaluate your strengths and any areas you need to revise	Practice test 1 solutions
❐	Complete Practice test 2 and hand in to your teacher	Practice test 2 for each test
❐	Ask your teacher for feedback on your performance and use of time	You and your teacher
❐	Keep updated on test information from your assessment authority	Test Administration Authority

Test day checklist

Use this checklist to ensure you are prepared for test day.

Day before the test

	Activity
❐	2B pencils, an eraser and a sharpener ready to take
❐	Have a good night's sleep

Test morning

	Activity
❐	Have breakfast
❐	Take watch, pencils, eraser and sharpener
❐	Arrive at school, or the testing venue, well before the session commences

During the test – Reading and Language Conventions

	Activity
❐	Be confident in your preparation
❐	Monitor your time during the test
❐	Work through the test, completing easy questions first
❐	Read each question carefully, underline important words
❐	Don't spend too much time on any one question
❐	Circle the question number of any question you need to return to
❐	Make sure your written answers are legible and show good understanding
❐	Only write in the box or on the lines provided
❐	Select the correct option(s) in a multiple-choice question; check the other options are incorrect
❐	Choose the correct number of answer options only (usually one) and colour the box(es) in completely
❐	If you change your answer, rub out the other answer completely
❐	Go back to complete unanswered questions
❐	Check your work, make sure you haven't skipped any questions

During the test – Writing

	Activity
❐	Know the particular genre being tested, for example narrative, recount
❐	Plan your writing, making a clear connection to the stimulus
❐	Meet genre expectations in your writing
❐	Keep in mind the marking criteria and how it is applied
❐	Only write on the lines provided and within the set length limitations
❐	Be mindful of the time, allow time to check your work
❐	Reread your writing and make any necessary revisions
❐	Check your spelling, grammar and punctuation

After the test

	Activity
❐	Discuss your results with your teacher
❐	Identify your strengths and any potential areas to revise
❐	Consider these results together with other evidence of your progress
❐	Review learning goals to ensure they are appropriate

Key Skills: Language Conventions

A convention is something that is expected to be done in a particular way and is an accepted practice. For example: when there is a queue, it is conventional to join the queue from the back and not the middle or front. Not following the convention may cause others to misinterpret our actions.

Without conventions in language we might end up with something we weren't expecting. Wrong spelling could have us eating a chocolate moose instead of a chocolate mousse for dessert.

There are conventions when we use language. Conventions for **spelling**, **grammar** and **punctuation** are used in English to ensure accuracy in expression.

These conventions help to ensure that, when we use language, our meaning is understood.

Spelling

The spelling section of the Language Conventions test will assess your ability to use accurate spelling. Some questions will ask you to identify the misspelt word and write the correct answer.

One type of question on the test states:

'Each sentence has one word that is incorrect.

Write the correct spelling of the word in the box.'

a) I left my shoes in the hall downstares. [ ]

The word that has been misspelt is: **downstares**. The correct spelling, **downstairs**, must be written in the box.

Another type of question to test your ability to spell will identify the spelling error for you and ask you to write the answer.

The question will state:

'The spelling mistakes in these texts have been circled.

Write the correct spelling for each circled word in the box.'

1) Snow although bootiful [ ]
and white can be
2) danjrous after big falls and [ ]
can prevent skiers from
3) acessing the higher [ ]
4) mountain peeks. [ ]

The correct answer for box 1) is 'beautiful'. The correct answer for box 2) is 'dangerous'. The correct answer for box 3) is 'accessing' and the correct answer for box 4) is 'peaks'.

Strategies for completing spelling questions

HOT TIP

1 Complete all the spelling questions you already know first and then go back and apply the following strategies to those questions you find more challenging.
2 **Review** spelling rules.
3 **Practice** identifying **misspelt words**.
4 **Highlight** the words that you know are commonly misspelt. For example:
I love choclate cake. [ ]
5 **Highlight** words where you know other letters could be used to make the same sounds in the words.
For example:
The line roared, frightening the crowd and forcing them to reel backwards. [ ]
6 **Highlight** any words that may look wrong.
7 **Write** the alternative options of the word you have identified as misspelt above the word in the sentence on the test paper.
For example:
received *speach* *nite*
The student receeved the award at Speech Night.
8 **Write** the word that fulfils the rule, looks right and sounds right in clear, legible handwriting in the box.

received

Punctuation

The Language Conventions test also contains questions about punctuation. Punctuation signs and symbols help to make the meaning in a sentence clear. If we did not use punctuation in the same way, the following sentence taken in part from a book by Lynne Truss could be misinterpreted.

The panda eats, shoots and leaves.

If the comma is inserted after 'eats', then the sentence means that after a panda eats his meal, he shoots, perhaps a weapon, and then leaves the scene of the crime. The writer probably doesn't mean this and without the comma after 'eats', the sentence tells the reader that pandas eat shoots of plants and leaves.

The Language Conventions test will assess your ability to use various conventions of punctuation by asking you to identify the correctly punctuated sentence. The question states:

'Which sentence has the correct punctuation?'

You then need to select the sentence which is accurately punctuated. You will need to shade one box which corresponds with the accurately punctuated sentence.

You may be asked about the punctuation for **direct speech**, **apostrophes**, **sentence endings,** or the use of **commas to separate clauses and verbless clauses**.

NOTE: In all cases you must shade the box which corresponds with the accurate response. Be careful, some sentences may appear to be correct at a quick glance; however, it is important you check the use of all punctuation in the sentence is correct. So check the sentence starts with a capital letter, ends with a full stop, question mark or exclamation mark, and that all the punctuation within the sentence such as capitals, apostrophes, direct speech quotations and commas are correct.

For example:

Which sentence has the correct punctuation?

☐ The Great Barrier Reef in Queensland is a popular tourist destination.

☐ The Great barrier Reef in Queensland is a popular tourist destination.

☐ the Great Barrier Reef in Queensland is a popular tourist destination.

☐ The Great Barrier Reef in Queensland is a popular tourist destination

The correct answer is the first sentence, as all capital letters are accurate and there is a full stop at the end of the sentence.

Another type of punctuation question will ask you to shade one box to indicate where an apostrophe should be in a word in a sentence.

The question states:

'Shade one box to show where the missing apostrophe (') should go.'

The cats☐ job is to chase the rats☐ until she catches☐ them.

Shade the box after the 't' in 'cats'. This is the correct response because there is one cat and she owns the job of catching rats.

Another type of punctuation question states:

Shade two boxes to show where the missing commas (,) should go.

This question is asking you to indicate how to separate a clause or verbless clause in a sentence. You must shade the box to show where the clause or verbless clause starts and where it finishes within the main sentence.

The girl☐ who always cheers enthusiastically for her☐ team☐ is absent.

Shade the box after 'girl' and 'team'. This is the correct response because the principal sentence is: 'The girl is absent.' Within the commas is: 'who always cheers enthusiastically for her team'. This is a clause giving extra detail about the girl. This is called an adjectival clause because it describes the noun (the girl) in the principal sentence.

9780170462198

Another type of punctuation question states:

'Which sentence has the correct punctuation?'

This sentence is asking you to shade the box next to the sentence which is most accurate and accurately punctuated.

- ☐ I said I wanted to have a go so I stood at the end of the queue
- ☐ I said, 'I wanted to have a go so I stood at the end of the queue.'
- ☐ I said I wanted to have a go, so I stood at the end of the queue.
- ☐ I said, 'I wanted to have a go,' so I stood at the end of the queue.

Shade the third box, 'I said I wanted to have a go, so I stood at the end of the queue.' The third sentence is correct because there are no directly spoken parts of the sentence. The sentence needs a comma to indicate the break between the two clauses and it ends in a full stop.

Strategies for completing punctuation questions

HOT TIP

For multiple-choice questions on punctuation, there is **NO** pattern in the options from which to choose. DO NOT ALWAYS CHOOSE THE FIRST RESPONSE.

1 **Review** the conventions for punctuation before the test.
2 **Read** the questions and instructions carefully. Punctuation and grammar questions are sometimes on the same page.
3 **Look** at what sort of error you are being asked to find.
4 With questions whose answers offer options that have full stops and words that must have capital letters, **choose** the answer that has the correct sentence ending [full stop (.) or question mark (?) or exclamation mark (!)] and the correct use of capital letters.
5 With questions whose answers offer options for direct speech, ensure the sentence needs quotation marks. Sometimes the sentence is in the past tense and no quotation marks are needed because there is no direct speech.
6 When answering questions that ask you to show where commas must be used around a clause or verbless clause, read each answer option carefully. **Identify** the main or principal sentence. This part of the sentence is essential. Then identify which parts of the sentence are added to offer the reader extra information about the main or principal sentence. Select the response which has a comma at the beginning and end of the extra information.
7 **Look** for when you take a breath or break in the sentence. This might give you a clue as to where the extra information is in the sentence.
8 In all questions for punctuation, **eliminate** any wrong answers and read your options carefully.
9 **Shade** the box which corresponds with the right answer.

Grammar

The Language Conventions test will also assess your ability to use grammar correctly. Using the conventions of grammar ensures that we express ourselves accurately, using the correct pronouns, tense of verbs, or the right adjective or adverb in a sentence.

Misplacing a prepositional phrase can be misleading. For example: 'A boy was in trouble today for running in the classroom with the principal.'

The writer does not mean that the boy and principal were running in the classroom. The details in the sentence need to be repositioned. It should say: 'A boy was in trouble with the principal today for running in the classroom.'

One type of grammar question states:

'Which of the following correctly completes the sentence?'

For this question, you will need to shade one box which corresponds with the correct tense of the verb for the sentence.

I have [ ] the dishes.

☐ wash ☐ washed ☐ washing ☐ washer

To answer the question, shade the correct form of the verb. The correct word for the space in the sentence is 'washed'.

Another type of question will have four similar sentences and you are required to shade the box which corresponds with the correct sentence. The question will state:

'Which sentence is correct?'

- ☐ The argument is between Sam and it.
- ☐ The argument is between Sam and me.
- ☐ The argument is between Sam and myself.
- ☐ The argument is between myself and Sam.

To correctly complete this question, shade the second box next to the sentence, 'The argument is between Sam and me'.

Strategies for completing grammar questions

HOT TIP

For multiple choice questions on grammar, there is **NO** pattern in the options from which to choose. DO NOT ALWAYS CHOOSE THE FIRST RESPONSE.

1 **Review** conventions of Grammar before the test.
2 **Read** the question and instructions carefully. Punctuation and grammar questions are sometimes on the same page.
3 **Do** all the questions you know and feel confident about the correct answer first.
4 With the more difficult questions go back, and **read** the sentence carefully.
5 For questions where you have to select an answer from four options, **eliminate** any options you know are incorrect.
6 When you are asked to select a word or words and insert into a sentence, **write** the options in the sentence on the test paper and read for accuracy. READ EVERY WORD IN A SENTENCE. Skimming over a sentence can lead to simple errors.
7 **Select** the option that sounds accurate and/or fulfils all the conventions you know and **shade** the box which corresponds with your choice.

Key terms: adjective, adverb, apostrophes, comma, convention, punctuation, tense. See the Glossary on page 107 for definitions of these terms.

9780170462198

TEST 1: Language Conventions

Instructions

- A correct answer scores 1 mark, and an incorrect answer scores 0.
- Marks are not deducted for incorrect answers.
- No marks will be given if more than the required number of boxes are shaded (most questions require one answer).
- Choose the alternative which most correctly answers the question and shade in the box next to it.

Spelling

WRITE YOUR OWN ANSWER

Each sentence has one word that is incorrect.
Write the correct spelling of the word in the box.

1 He actualy ran quite slowly for a tall man.

2 I keep all my treasures in a box underneeth my bed.

3 The final clue was used to solve the mistery.

4 Despite his bad reputation, the man was agreable and friendly.

5 She always carrys her handbag whenever she intends to go shopping.

6 She gose to hockey practice once a week.

7 I left the dore open so the dog could follow me inside.

The spelling mistakes in the following text have been circled.
Write the correct spelling for each circled word in the box.

WRITE YOUR OWN ANSWER

8 We all thought the girl was a courragous

surfer when she batled the huge

swell to cach the

perfect wave back to sure.

Each sentence has one word that is incorrect.
Write the correct spelling of the word in the box.

WRITE YOUR OWN ANSWER

9 It was an imagnative story filled with tales of fairies, magical castles and monsters.

10 My favourite styles of mewsic are rock and classical.

11 Over their, beneath the trees, is where the magic key is hidden.

12 Swiming is a popular spectator sport of the Olympic Games.

13 When I thaught about it, I realised that I had acted too hastily.

14 I spent an enjoyible afternoon catching up with friends.

15 The ninteenth track on the album is the most popular.

The spelling mistakes in the following text have been circled. Write the correct spelling for each circled word in the box.

WRITE YOUR OWN ANSWER

16 I recieved praise from the head chef

when she saw that I had: seperated

the eggs corectly, measured and added

each ingrediant and created the mixture

for the eigth and last Christmas cake.

Grammar

SHADE ONE BOX

1 Which of the following correctly completes the sentence?

He was the ☐ competitor in the race.

☐ fast ☐ fastest ☐ most fastest ☐ faster

2 Which of the following correctly completes the sentence?

SHADE ONE BOX

I ☐ been here earlier, but the traffic was slow.

☐ would'ave ☐ wouldve ☐ would've ☐ would of

9780170462198

3 Which of the following correctly completes the sentence?

SHADE ONE BOX

He thought he ☐ his most valuable possession.

☐ has lost ☐ will lost ☐ loses ☐ had lost

4 Which sentence is correct?

SHADE ONE BOX

☐ She was calling him on the phone and doing her homework at the same time.

☐ She was called him on the phone and doing her homework at the same time.

☐ She was calling him on the phone and done her homework at the same time.

☐ She was called him on the phone and had done her homework at the same time.

5 Which sentence is correct?

SHADE ONE BOX

☐ The camp leaders are choosing between Max and me.

☐ The camp leaders are choosing between Max and him.

☐ The camp leaders are choosing between Max and myself.

☐ The camp leaders are choosing between myself and Max.

6 Which of the following correctly completes the sentence?

SHADE ONE BOX

My favourite meal is lunch, especially when it includes a ☐ piece of fruit.

☐ exciting ☐ pretty ☐ attractive ☐ tasty

7 Which of the following correctly completes the sentence?

SHADE ONE BOX

I ☐ to finish my chores before I can play.

☐ had ☐ having ☐ have ☐ haved

8 Which of the following correctly completes the sentence?

SHADE ONE BOX

Even though I arrived late, I [ ] played in the second half.

☐ could of ☐ could've ☐ could'ave ☐ couldve

9 Which of the following correctly completes the sentence?

SHADE ONE BOX

The dog [ ] his bone in our yard yesterday.

☐ has buried ☐ buries ☐ will bury ☐ had buried

10 Which sentence is correct?

SHADE ONE BOX

☐ They were honours at the ceremony and presented a medal.
☐ They were honoured at the ceremony and presents a medal.
☐ They were honoured at the ceremony and presented a medal.
☐ They were honour at the ceremony and presented a medal.

11 Which sentence is correct?

SHADE ONE BOX

☐ The argument is between Sally and me.
☐ The argument is between Sally and it.
☐ The argument is between Sally and myself.
☐ The argument is between I and Sally.

12 Which of the following correctly completes the sentence?

SHADE ONE BOX

The new park has [ ] gym equipment to encourage the local community to be more active and exercise outdoors.

☐ lovely ☐ useful ☐ gorgeous ☐ delightful

Punctuation

SHADE ONE BOX

1 Which sentence has the correct punctuation?

☐ The habitats of Australia's native animals are threatened by introduced species
☐ The habitats of australia's native animals are threatened by introduced species.
☐ The habitats of Australias native animals are threatened by introduced species.
☐ The habitats of Australia's native animals are threatened by introduced species.

2 Which sentence has the correct punctuation?

SHADE ONE BOX

☐ She said to, 'bring a present for Olivia.'

☐ She said, 'to bring a present for Olivia.'

☐ She said to bring a present for Olivia.

☐ She said to, bring a present for Olivia.

3 Which sentence has the correct punctuation?

SHADE ONE BOX

☐ He said, 'Run to the corner.'

☐ He said, 'run to the corner.'

☐ He said run to the corner.

☐ He said, 'run,' to the corner.

4 Which sentence has the correct punctuation?

SHADE ONE BOX

☐ The large, shells even those that have been chipped, were once homes for sea creatures.

☐ The large shells, even those that have been chipped, were once homes for sea creatures.

☐ The large shells even those that have been chipped, were once homes, for sea creatures.

☐ The large shells even those that have been chipped were once homes for sea creatures.

5 Shade one box to show where the missing apostrophe (') should go.

SHADE ONE BOX

The baby polar bear ☐ enclosure is one of the zoos ☐ most ☐ popular exhibits ☐.

6 Shade **two** boxes to show where the missing commas (,) should go.

SHADE TWO BOXES

Uluru ☐ in the ☐ Northern Territory ☐ is often photographed ☐ at sunset.

7 Which sentence has the correct punctuation?

SHADE ONE BOX

☐ Laura's birthday is in January.

☐ Lauras birthday is in January.

☐ Laura's birthday is in January

☐ Lauras' birthday is in January.

8 Which sentence has the correct punctuation?

SHADE ONE BOX

- [ ] He said, 'we had missed the movie because we arrived late.'
- [ ] He said, 'we had missed the movie,' because we arrived late.
- [ ] He said we had missed the movie, because we arrived late.
- [ ] He said, we had missed the movie, because we arrived late.

9 Which sentence has the correct punctuation?

SHADE ONE BOX

- [ ] I can reach it myself, said Hamish.
- [ ] 'I can reach it myself,' said Hamish.
- [ ] I can reach it myself 'said Hamish.'
- [ ] 'I can reach it myself,' Said Hamish.

10 Which sentence has the correct punctuation?

SHADE ONE BOX

- [ ] My youngest cousin, William who loves cars, is going to be at the party.
- [ ] My youngest cousin William, who loves cars, is going to be at the party.
- [ ] My youngest, cousin William who loves cars, is going to be at the party.
- [ ] My youngest cousin William who loves cars is going, to be, at the party.

11 Shade one box to show where the missing apostrophe (') should go.

SHADE ONE BOX

The trains passengers were students and shift workers.

12 Shade **two** boxes to show where the missing commas (,) should go.

SHADE TWO BOXES

The old hotel that had provided accommodation to so many weary travellers was to be relocated because of the new road.

9780170462198

Reading – what's it all about?

Almost everybody can learn to read by recognising meaning in signs, symbols and the alphabet. Understanding how meaning is created, when we read, takes a greater skill than simple recognition. We read for different reasons and the types of texts we read can provide enjoyment, information, knowledge, etc. We can skim and scan to get a general overview; read the whole text; analyse and deconstruct texts to comprehend meaning. Good readers are able to locate, comprehend and extract information from a variety of texts.

Reading is an essential part of our everyday lives; it is something we do every day without realising that we are actually reading. Reading helps us to make sense of things and this can be visual, through images, films, and photographs; written through printed texts; and the multimedia literacies of today's world. Reading is now an important element of social networking and instantly connects us with the rest of the world.

What happens when we read?

Reading is not done in isolation. It is not a chore to be endured. The purpose of reading is to construct meaning *from* a text. Reading is more than decoding the signs and symbols of the text: it is an interactive process that occurs between the reader and the text, resulting in the understanding of meaning. This occurs when we know which language features are appropriate for a specific type of text. Comprehension is when we demonstrate the knowledge and skills used to understand the text through appropriate responses to questioning about the text. This knowledge and these skills also assist our interaction with other language use (writing, speaking, listening, representing and viewing). Those skills are transferable to other texts when we understand the content and context juxtaposed with the interactive process of reading.

Reading is a combination of knowledge and skills that helps us to decode the meaning in a text. When we read, our brain makes sense of the 'ink blots' on a page at lightning speed and works out what the text means. This process is the same regardless of the way a text is published. There are four aspects in reading proficiently.

1 Language conventions are the functional part of the text. Letter/sound combinations, grammar and syntax, word/sentences/paragraph structure, spelling and punctuation combine to establish the meaning in the text.
2 Comprehending or understanding how the ideas in a text relate to each other. At this level we are interacting with the text. Sometimes something we already know can help us to understand other texts. As our reading skills develop we can use the knowledge learnt to help us understand more difficult texts. There are two basic levels of comprehending: literal and inferential.
3 Purpose is working out why a text was composed and how it relates to us. At this level we consider how and why the purpose of the text has shaped the way it has been composed. The purpose shapes the structure, tone, formality and sequence and helps us understand the layers of meaning in the text. Texts can be composed for different purposes.
4 Evaluate, identify and analyse ways in which ideas are expressed to influence reading perceptions. We improve our understanding of how texts are crafted by exploring the values expressed in texts. What is the text asking me to believe? How does the composer use language to achieve this purpose? At this level we express our own views on the meaning of the text. This can be to agree or disagree with the composer's point of view. We interact with the text on both emotional and intellectual levels.

Three levels of understanding

Reading to understand meaning encourages three layers of questioning. These are referred to as ***here***, ***hidden*** and ***head***.

Here: the right answer is in the text as either a word or a sentence. The composer said it. We do not have to think as hard to find a *here* answer.

Hidden: the right answer requires some interpretation or searching for information. The composer meant it. It is found by joining together information from two or more places in the text, or information from the text and what you already know.

Head: the right answer requires your own evaluation or understanding that goes beyond the literal text. The composer would agree with it. It comes from your background knowledge of the text. What is it about and how will this help you to puzzle out the answer in your head?

Each level of questioning, literal, interpretive and evaluative, demonstrates understanding of layers of meaning which are inherent in texts.

In the Reading test you will find questions that ask you to find answers from each of the above three levels. As you work your way through the questions it is important to consider which of the three levels the question is asking.

To become a more efficient reader you need to be actively participating in the whole text.

1 Before reading investigate the text in front of you. The title is important. How is the text structured and/or organised? Try to predict what it may be about through the cover and the title. What knowledge do you already have that will be important in understanding this text?
2 As you read, involve yourself in the text. Improving your reading skills involves extending your vocabulary with unknown words. Using context clues, for example drawing maps or pictures, may help your understanding. Reread a section if you didn't understand it the first time. Try to link back to what you already knew or predicted in the prereading and note how that can also provide clues. Make notes as you read about what you understand as important details. If you have questions can you answer them from the text or will you need to find extra information, outside the text?
3 After you read combine what you knew with what you have learnt. You have improved your reading skills! Summarising the main ideas will consolidate your understanding. Mind maps or graphic organisers are useful in reviewing the main points.

Simple to complex

The questions you will be asked to answer will move from questions of basic comprehension to increasingly complex interpretive and inferential questions.

For example, each text will begin with a *here* comprehension question: a 'where' or 'what' question and phrases like 'According to the text . . ., 'The words __________ tell us that . . .', 'This text is mainly about the __________ . . .' These are general questions.

The questions will become more challenging. This can include the language conventions within the text and the type of language techniques, for example descriptive, emotive and figurative. The type of language used will support the composer's purpose. These questions are at the *hidden* level. These are specific questions that ask you to connect information. A riddle is an example of connecting information to find an answer.

The most challenging questions ask you to interpret and infer, and require your brain to work much harder to determine the answer. These are the 'how' or 'why' questions. For example 'How do you know the boy likes soccer?' Interpreting the effect resulting from a composer's use of a particular writing style, the techniques they use, and selecting specific information requires higher order thinking skills.

Skills and understandings you need

1 Scanning is the way we read a telephone book or television guide. Our eyes move quickly looking for a particular word or words that are relevant. It is not necessary to read every word. There is no main idea or theme to discover.
2 Skimming is when we read quickly through a text to get the main idea or theme by reading headings, summaries, first and last sentence of each paragraph rather than reading the whole text.
3 Topic sentences are usually, but not always, at the beginning of a paragraph. Identifying the topic sentence will help with summarising and note taking.
4 Reading in detail implies you are reading every word. This is important when gathering information to ensure 100% accuracy is achieved before applying that information. **More time is spent actually understanding the text when it is read, therefore deepening your understanding of the text.**
5 Make a note of key words as you read. You can do this in point form with one or two key words for each point.
6 Vocabulary should always be considered in the context of the text. Check for a glossary of words at the back of your text. The dictionary definition may not be the precise meaning of the word in context, for example, technical words.

You can mix these strategies depending on your purpose for reading. Scanning and skimming as an introduction followed by reading in detail is effective.

Understanding features of different text types

All texts are constructed. The different structures are referred to as text types. Knowing the difference helps us to recognise that the language conventions, layout or structure, and purpose are distinctive for the text type. The ability to visualise (in our head) the features of a text type, and how those features are arranged, is essential to understanding the constructed meaning, when we are reading.

Narrative

The basic purpose of a narrative text is to entertain through story-telling, which grabs and holds the reader's interest. Narratives can be found everywhere.

Novels, novellas, short stories, films, poems, paintings, photographs, soap operas and song lyrics are some examples. They can be fiction or factual or a combination of both. A narrative structure has an introduction where we are orientated to background information about who, where and when. This is followed by a complication or problem which starts a series of events that are finally resolved. The story is often told through a series of events in the order they happen.

The features common across all narratives are characters with obvious personalities, physical descriptions and identities. Dialogue between the characters occurs on a regular basis to help develop the plot. Language is descriptive to increase the intensity of the narrative and maintain the interest of the reader. The composer will ask rhetorical questions to provoke a response from the reader. Setting is the time and place of the action. Tense can shift from past to present to future; however, past tense is used more often.

Recount

Recounts are composed to inform, entertain or explain past events. These events are retold in chronological order. There are some similarities to narrative texts but the differences are distinct and easily understood. Recounts have an orientation and a sequence of events. The conclusion is a summary and often expresses personal opinions of the composer. It is always written in past tense.

There are three types of recounts. Factual recounts such as the television news communicate events, usually shortly after those events, to local and international audiences. Personal recounts tell us about a family holiday, outing or function and this is a more social text. Explaining how a task was completed is a procedural recount and it is found in science and hospitality subjects. Other examples of recount texts are letters, eyewitness accounts, dramatic re-enactments diaries and journals. Social online networking such as blogs and communities are also examples of recount text types.

Language is less descriptive and more factual, technical and usually less emotive than a narrative text. However, descriptive language is acceptable too. Proper nouns are used for names of people and places. Time and sequence words, such as 'then, 'next' 'finally', are used to show the order of the events. Cause and effect words, such as 'because of', 'in order to', or 'as a result', are also important in connecting the events.

Exposition

An exposition is a text of argument. It debates one side of a topic or question and presents a singular side of an argument through a predetermined point of view. The composer attempts to persuade the reader to argee with the point of view presented, for example in debates, legal defences, editorials and arguments, or to do something or act in a certain way, for example advertising.

The three parts of an exposition are:

- an introductory statement that presents the writer's point of view and previews the arguments to be presented
- a number of arguments that aim to persuade the reader. A new paragraph is used for each new argument
- a conclusion that sums up those arguments and reinforces the point of view.

Advertising aims to convince us to buy or use a particular product or service. Visual elements are important in persuading the reader and must always be considered as one element within the whole text.

Language used is emotive and shows us feelings and attitudes in support of a point of view. Word choice is formal and persuasive indicating knowledge of the subject of the argument. Present tense is always used to describe current issues. Cause and effect words such as 'otherwise', 'because', 'consequently', or 'leads to' and linking words such as 'in addition', 'also', 'moreover' and 'as well' help to develop a cohesive argument. Including references to research, statistics and experts adds power to an exposition text.

Information

The purpose of an information text is to provide accurate, non-fiction reports or descriptions. The composer is usually an expert who is able to provide detailed information, for example scientific and environmental information about the natural and physical world. Informative texts will also include diagrams, photographs, tables, illustrations and graphs as supporting evidence. The type of information, which can be presented, is very broad and ranges from a letter advising of a school excursion to the latest information received from NASA's space tracking. Each is of equal importance for a different purpose and audience. Informative texts are prevalent in our everyday lives and provide us with much of the day-to-day factual information we need.

Language is specific and/or technical to the information presented. Emotive or descriptive language is minimised. Present tense and third person is used. Sentences are simple or compound to maintain clear meaning. Complex sentences are used for cause and effect.

The three parts of an information text are:

- an introductory statement
- a series of facts, supported by evidence and visual texts, about various features of the topic. These

facts are clearly organised in paragraphs
- a conclusion that sums up the contents of the text.

A bibliography, which contains a list of the sources used to research the information, is included at the end of the text.

Key terms: cohesion, complex sentence, composer, compound sentence, context, convention, deconstruct, dialogue, exposition, inference, informative, narrative, recount, rhetorical question, scan, sequence, skim, structure, symbol, tense, topic sentence. See the Glossary on page 107 for definitions of these terms.

TEST 2: Reading

Instructions

- A correct answer scores 1 mark, and an incorrect answer scores 0.
- Marks are not deducted for incorrect answers.
- No marks are given if more than one answer alternative is shaded.
- Choose the alternative which most correctly answers the question and shade in the box next to it.

Section A: Narrative

Read the following narrative and then answer the questions that follow.

Chewy Under the Table

Steven Cumper

Morgan was seven years old and he noticed things. He'd just had a baby sister arrive and the house was a mess. His Mum and Dad were yelling a lot […]

On Sundays his family would go over to Grandma and Grandpa's house to help out with chores and stuff. After this, Dad, Grandpa and Uncle Frank shouted at the footy match on TV. Mum was kept busy with baby Charlotte, and Morgan didn't think she minded being left alone […]

Morgan noticed that Grandma had curly hair, wore a faded floral pinnie and smelled of Lux and sweat. He would help her on these mornings and prattled away whilst she nodded and chopped, grated, stirred and sliced. The little kitchen was bustling and the air was filled with the sweet smell of stewing Granny Smiths and the expectation of a yummy roast. Morgan watched fascinated as his Grandma grunted and groaned while she laboriously rolled pastry, her upper arms wobbling furiously where her muscles should have been […]

The table was set, complete with a jug of steaming Gravox and a haphazard stack of thickly buttered white bread. Everybody hoed into the roast with a vigorous salting and mopping of gravy […] Mum and baby Charlotte retired early to the lounge to watch the muscled, sweaty men on Epic Theatre. Grandma cleared up and, with a twang of the screen door, went outside to listen to the trannie and the tweeting of Mr Chips, the budgie, on the porch. Morgan retreated quietly to his spot under the table.

The hairy legs of his Dad, Grandpa and Uncle Frank looked skinny and funny down there. It was getting really loud at the table and the men's voices had changed. They were becoming more and more excited, and Morgan felt the table being thumped menacingly several times. […] It was usually about this time that he noticed the little wads of dried chewing gum poked into the corners of the kitchen table. There must have been 'hundreds of 'em', he thought, mouthing the words quietly to himself over the noise above. He began to count them, like he had the week before and the week before that, each time finding a couple more new, still soft ones […]

Soon it would be over and Grandma's soft, furry kiss would signal the time to leave. Grandpa had gone off for a rest and Uncle Frank helped Dad into the car. With his Mum's jaw set grimly, they drove toward home at dusk. Baby Charlotte was lying quietly in the bassinet on the back seat and his dad was snoring heavily in the front. Their stationwagon glided through the darkening streets. Morgan looked over to his Mum behind the wheel and thought about telling her. Instead, he pressed his face to the cold window and exhaled expanding breath clouds onto it before drawing smiley pictures with his fingers. 'Dragon's breath,' his dad would call it. He gazed into the windows of passing strangers' homes that momentarily flashed their warm, yellow glow across his face. He wanted to tell his mum that today he had counted to 273.

Shade the correct box to answer the following questions:

1 Morgan's house was a mess because

SHADE ONE BOX

- ☐ he noticed things.
- ☐ his mother and father were in a hurry as they were going out.
- ☐ he had a new baby in the family.
- ☐ he was seven years old.

2 Morgan's family went to visit his Grandma and Grandpa's house on Sunday so

SHADE ONE BOX

- ☐ they could help do some jobs.
- ☐ they could have a big roast dinner.
- ☐ Grandma could take care of Charlotte.
- ☐ they could watch the football.

3 'The little kitchen was bustling' tells us

SHADE ONE BOX

- ☐ the house was on a busy road.
- ☐ a lot of activity was taking place.
- ☐ Morgan was playing with his toys.
- ☐ it was an interesting place to be.

4 'Everybody hoed into the roast' means

SHADE ONE BOX

- ☐ Grandma was not a very good cook.
- ☐ Morgan enjoyed Sundays.
- ☐ everyone ate heartily and fast.
- ☐ the family enjoyed being together.

5 Morgan usually noticed the chewing gum under the table

SHADE ONE BOX

- ☐ while he was watching the football on television.
- ☐ as soon as he got under the table.
- ☐ not until the men's voices got louder.
- ☐ when his Grandma went out on the porch.

6 Morgan thought about telling his mother that

SHADE ONE BOX

- ☐ his father was snoring too loudly.
- ☐ Charlotte was asleep in the backseat.
- ☐ he had drawn smiley faces on the car window.
- ☐ he had counted a lot of pieces of gum today.

Section B: Recount

Read the following recount and then answer the questions that follow.

The Cat Show

Last weekend I took my two Maine Coon cats, Puzzle and Kynan, to a cat show. It was held near the Blue Mountains in New South Wales. There were many different breeds of cats and kittens. This cat show was a great place to learn about having cats as pets. The cat was first domesticated about 5,000 years ago in Ancient Egypt so people have had them as pets for a very long time.

Preparations began several days before so that my cats would be in top show condition. Everyday grooming they did for themselves but preparing for a show takes extra human help. Cat shows are beauty competitions with set standards for different breeds.

The first thing I did was clip their claws because cats that live indoors are not able to do their own manicures on tree trunks. It was important to do this so my cats did not accidentally scratch a judge.

The next step was to give them a bath; luckily my cats enjoyed having a bath. I used special cat shampoo and conditioner that makes their fur soft and smell nice. After their baths Puzzle and Kynan needed to be brushed to remove any loose fur and matting.

When we arrived my cats were checked by a vet. I was then given the number for their cages. All the cages were set out on tables and in rows in a big hall. After I put curtains in the cages and gave them a quick brush they were settled and ready for the judging.

It took about two hours for the judging. The judges took the cats out, picked them up and held them in order to award points. After the judges finished looking at all the cats they had to go away and add up all the points and work out which cat was the best of their breed. Once they had finished then they worked out which cat was the very best cat for the day from all the breeds.

Puzzle and Kynan had a good day and won lots of ribbons and prizes. They also enjoyed the children who stopped and said hello throughout the day. We were all tired and had a long drive back; they slept all the way home.

9780170462198

Shade the correct box to answer the following questions:

1 'The cat was first domesticated about 5,000 years ago . . .'

SHADE ONE BOX

This tells us that cats

- ☐ were first found in Eygpt.
- ☐ had very high standards.
- ☐ had limited breeds.
- ☐ were kept as pets.

2 According to the text, Puzzle and Kynan had a bath

SHADE ONE BOX

- ☐ because they can't wash themselves.
- ☐ because cats are dirty animals.
- ☐ to prepare their fur for clipping.
- ☐ to prepare their coats to impress the judges.

3 According to the text cats have their claws clipped before a show because

SHADE ONE BOX

- ☐ they can't do it for themselves.
- ☐ the nails grow too quickly for the cat to manage.
- ☐ they might accidentally injure a judge.
- ☐ they can't go to the nail shop for a manicure.

4 What is the most important point/message in this story?

SHADE ONE BOX

- ☐ The cats enjoy the experience and all the attention.
- ☐ The judges get to compare lots of different cats.
- ☐ Cats are important throughout history.
- ☐ It is a very tiring day and a long distance to travel.

Section C: Expository

Read the following expository text and then answer the questions that follow.

TravelSmart Snapshots

TravelSmart Snapshots
strong increases in public transport patronage
making it easy to find local services
cheaper and more sustainable ways of getting around

Western Australia

In Western Australia, *TravelSmart Household*, a personalised travel information service, helps the community make better use of the available travel options by replacing car trips with walking, cycling and public transport.

This approach, pioneered in Western Australia, is now informing similar approaches throughout Australia, North America and Europe. It proves that car use (km driven) can be reduced by around 13 per cent.

The WA *TravelSmart Household* program involves 258,000 residents across seventeen metropolitan communities. The project resulted in strong increases in public transport patronage, annual reductions achieved in the first eight communities of 100 million car kilometres (equivalent to taking 6,600 cars off the road) and saving 30,000 tonnes of greenhouse gas.

Victoria

TravelSmart Victoria is running the largest multi-modal household travel behaviour change program in the world. Working in the western suburbs of the Cities of Maribyrnong and Moonee Valley, the program will reach 50,000 households.

In addition, Victoria is testing new travel behaviour change methodologies for schools and workplaces in congested areas in the Melbourne metropolitan area. These include new car pooling software, a project promoting walking and cycling through the use of pedometers and cycle speedometers and software allowing employers to calculate the dollar costs how their staff travel. For schools, the program is developing and testing a guide to producing school travel plans.

In 2005-06, *TravelSmart* will seek to develop green travel plans for 50 large employers in the Melbourne CBD as part of an effort to alleviate traffic congestion expected with the 2006 Commonwealth Games.

Queensland

Queensland implements the full range of voluntary *TravelSmart* programs in schools, workplaces, communities and destinations.

TravelSmart Destination projects are among some of Queensland's most successful projects. Large destinations such as universities are well positioned to encourage students, staff and visitors to minimize car trips.

James Cook University in Townsville and the Queensland University of Technology's Kelvin Grove campus have been the first to take the lead in helping to reduce traffic, improve health and preserve the environment.

James Cook University reduced emissions and achieved a 20 per cent reduction in car use. Queensland University of Technology achieved a 9 percent increase in train use, and a 16 percent increase in bus use. There has also been an 18.5 percent increase in walking and cycling. The strategies to achieve these results included a range of promotional activities and policy initiatives.

New South Wales

The first *TravelSmart Households* pilot program in NSW has taken place recently in Ermington and Woy Woy. About 5,600 households were invited to participate in a program designed to encourage residents to leave the car at home. Following the program's evaluation, results will be published and future directions considered.

NSW also assists trip generators to produce transport access guides that provide customised sustainable transport information for people travelling to and from a particular site.

The University of Newcastle developed a comprehensive transport access plan for the Central Coast Campus at Ourimbah. As a result, the University is implementing improved travel information and transport services, improved parking management, as well as infrastructure improvements for cyclists and pedestrians.

South Australia

TravelSmart in South Australia is working closely with other government agencies, the Australian Greenhouse Office and primary schools to reduce transport-related greenhouse emissions.

The *Green Travel Challenge* aims to increase the number of primary school students using sustainable modes of transport for trips to school and engage primary schools in activities that educate and challenge students to walk, cycle or car pool to school.

The *Green Travel Challenge* is delivered to schools via a CD-Rom package or on-line on the Transport SA website.

It is anticipated that approximately 1,000 students will complete the challenge by July 2005 from schools within South Australia.

An extensive program is also active in South Australia to support households and communities to make positive changes in their travel patterns.

Australian Capital Territory

In Canberra, *TraveSmart* is an integral component of the ACT's *Sustainable Transport Plan*. The ACT's *TravelSmart* program consists of *TravelSmart Workplaces* and *TravelSmart Households*.

TravelSmart Workplaces involves developing customised Travel Plans for five workplaces to encourage staff to leave their cars at home and use healthier, more sustainable modes of transport as a means of travelling to and from work. Some innovative ideas such as car-pooling, working from home, teleconferencing are suggested. Staff, employers and the local neighbourhood are expected to realize many benefits once the plans have been implemented.

A pilot study, *Households on the Move*, was conducted to test the idea that moving house is an opportune time to establish healthier, more sustainable travel habits. Following this, a large scale *TravelSmart Households* program is being developed targeting at least 11,000 households.

TravelSmart Australia
a better way to go

9780170462198

Shade the correct box to answer the following questions:

SHADE ONE BOX

1 The purpose of TravelSmart Snapshots is to

- ☐ make life easier for everyone.
- ☐ make people walk to get healthy.
- ☐ encourage students to wear hats.
- ☐ promote awareness of alternative transport.

2 How many Western Australians are involved in this project?

SHADE ONE BOX

☐ 6600 ☐ 30 000 ☐ 258 000 ☐ 100 million

3 Which definition best describes the program?

SHADE ONE BOX

☐ more buses ☐ innovative ☐ snapshot ☐ public service

4 According to the text, which State or Territory is developing the ideas of car-pooling and using the opportunity of moving house to start better habits?

SHADE ONE BOX

- ☐ Queensland
- ☐ New South Wales
- ☐ South Australia
- ☐ Australian Capital Territory

5 How is the 'Green Travel Challenge' delivered to schools in South Australia?

SHADE ONE BOX

- ☐ electronic media
- ☐ local television
- ☐ school newsletters
- ☐ community meetings

Section D: Informative

Read the following informative text and then answer the questions that follow.

Shark Tales Inspired Terror Through the Ages

The Daily Telegraph Tuesday Jan 13, 2009 p35.
These sleek fish have intrigued for centuries.
Troy Lennon

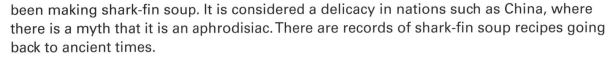

Sharks were here millions of years before people. But as soon as humans first came into contact with elasmobranchii, the two species developed a love-hate relationship. Many sharks are dangerous predators, yet some are gentle. Even those that are dangerous at least earn the respect of people who make their living on the ocean and some even became important resources.

In ancient Greece, legends tell of shark-like sea monsters devouring those who strayed too near their demesne.

Shark myths are also prevalent in Asia. For thousands of years Asians have been making shark-fin soup. It is considered a delicacy in nations such as China, where there is a myth that it is an aphrodisiac. There are records of shark-fin soup recipes going back to ancient times.

The soup is only made from the fin, many shark fisherman catch the animal, cut off its fin and then leave it to die. Many countries have banned shark finning but the practice continues.

When Melanesian and Polynesian people spread out across the Pacific over several centuries from about 1300BC, they came to fear the shark and even, in some cases, respect and worship it.

Australian Aborigines also revered sharks. Sharks' teeth were used for spear tips and jewellery, the skin could be used like sandpaper and nutrients could be obtained from shark meat. But the shark is also the animal totem for many clans.

Meriam Islanders or Murray people tell a legend of a father and son being stranded in the sea without their canoe. Sharks brushed the feet of the pair but did not harm them because the creature was their totem and it protected them. The name of the one clan on the island is Shark.

Aborigines in the Northern Territory tell of Mana, a shark ancestor who named many of the natural features of the coast, created their homeland and gave birth to all of the sharks in the ocean. Another Dreamtime story tells of Bul Mandji, who is killed by ancestral warrior Murayana using a special magic stingray spear. Bul Mandji is then carved up and the parts of his body become the ancestral lands of various people.

For some Indigenous Australians the constellation of the Southern Cross, only visible at night, shows a shark that is chasing a stingray across the sky.

Shade the correct box to answer the following questions:

SHADE ONE BOX

1 The main impression this text gives is that humans

☐ have little respect for sharks.

☐ think sharks can do no wrong.

☐ have an ambivalent relationship with sharks.

☐ view sharks as solely evil and troublesome.

---

2 The purpose of this text is to

SHADE ONE BOX

☐ tell interesting stories about the relationship between people and sharks.

☐ provide scientific research about sharks.

☐ promote shark conservation.

☐ dispel myths about sharks.

---

3 'These sleek fish have intrigued for centuries.'
What does 'these sleek fish' mean?

SHADE ONE BOX

☐ That sharks have smooth bodies.

☐ That sharks are fascinating.

☐ That sharks are dangerous.

☐ That sharks live in the sea.

---

4 According to the text, 'shark finning' is

SHADE ONE BOX

☐ cruel.

☐ banned in some countries.

☐ an ancient name for catching sharks.

☐ essential to make shark fin soup.

---

5 According to the text, 'Australian Aborigines also revered sharks'.
What does the word 'revered' mean?

SHADE ONE BOX

☐ They were afraid of sharks.

☐ They made shark fin soup.

☐ They respected and admired the sharks.

☐ They often kept sharks as pets.

---

6 'The constellation of the Southern Cross' is

SHADE ONE BOX

☐ jewellery made from sharks' teeth.

☐ a formation of stars.

☐ an Aboriginal artwork.

☐ the ocean where sharks live.

7 Give one reason why sharks are 'intriguing'. Use evidence from the text to support your answer, and write your answer below.

WRITE YOUR OWN ANSWER

9780170462198

Key Skills: Writing

The writing task is the same for Years 3, 5, 7 and 9 and it is marked using the same rubric for all students, regardless of year level. The task will specify the type of text that is required and has a prompt or stimulus which usually includes images to provide a springboard into the task.

Writing is a craft to be practised, edited and developed to make it interesting and enjoyable to read. The function of writing is to compose a range of literary and factual texts for a variety of purposes and topics for reader enjoyment and understanding. The craft of writing is developed through the effective use of language, which develops meaning, purpose and interest for an audience using a clearly identifiable text form.

You are required to develop a cohesive piece of writing. This means that each of the elements or components must be joined together in a logical and precise way, in the correct form, to produce a sophisticated and engaging text.

There are a number of text types that provide the basic structure for all forms of writing and each is written for a specific purpose and audience.

Literary texts	**Factual texts**
Drama	Description
Narrative	Discussion
Poetry	Explanation
	Exposition
Narrative texts in a wider sense can be factual as well as literary and include biographies, recount, procedural recount, procedure, report, response or review (book, film, game, website etc.)	

Writing well involves a good vocabulary. The first step for increasing your word choices is to read widely. The repetitious use of common words is a very widespread mistake. Reading not only introduces you to new words but also provides contexts for using those words correctly.

Another way to build your vocabulary is by using a thesaurus. However, be wary of taking the thesaurus too literally; a word listed as a synonym may not be a perfect replacement for the word you are already using. You will not have access to a thesaurus during the test, therefore you need to develop your word choices as you learn about the craft of writing.

The essential components in developing this craft, to ensure it is cohesive and enjoyable for the reader, include:

- spelling
- punctuation
- paragraphs
- modality
- grammar
- sentences
- word choice
- context
- linking
- engage an audience
- textual integrity
- consistency
- appropriate vocabulary
- range of literary devices
- purpose and audience
- text type features.

Sequencing correctly is essential when writing in any genre (for example: romance, science fiction, crime, etc.). Being able to write in any genre requires skill and understanding of those genres. Your writing must maintain a structure that allows the reader to easily follow and understand what you have written. Structure and sequencing, so the ideas will flow into each other, requires you to stay on topic and use well controlled language. Excellent writers understand and develop language choices similar to those used in different genres through the process of wide reading. Consider the type of language necessary for different text types.

A good writer also continues to develop and extend their skills to become a better writer. Always edit your own work by checking for repeated basic vocabulary and spelling errors. Also check that sentences and paragraphs are correct, tense is consistent and that your overall meaning flows and connects in a logical sequence.

Types of texts

A **narrative** has a specific structure, which includes an introduction to characters and setting; followed by an everyday event that may start to go wrong; followed by a series of events that finally arrive at some resolution. Language used in narrative writing is descriptive and helps us to understand the feelings and emotions of the characters. Effective word combinations include similes and metaphors (comparisons which create visual images), noun groups, strong action verbs and adverbs as well as correct tense and pronoun referencing.

A **recount** is an information text that retells or recalls an event. The introduction tells us who, what, where and when in the correct chronological sequence. Language used must connect time to logically develop the sequence. Effective word combinations include 'when', 'then', 'later', 'the next day', 'after', 'finally'. Avoid overuse of the word 'then'. Explain how the thoughts and feelings of the participants in the events affect their actions. Provide details without unnecessary descriptions. The ending provides a brief summary.

An **exposition** is used to develop an argument, which presents a point of view in a logical order. It argues a case for or against a particular position. Examples of exposition texts can include debates, editorials, arguments and advertisements. You should state your position in the introduction and develop reasons for your point of view. Language of argument or persuasion should be used. Present tense verbs, which carry emotion, are used. Your argument should be justified. An exposition does not discuss both sides of an argument; it is essential that only one side be argued. The ending should summarise and restate your point of view in relation to the argument.

An **informative** text is a non-fiction text, which presents factual information to report or describe something. Newspapers are one example of information texts although information texts are found everywhere. Language used is technical rather than emotional. Sentences are simple or compound to keep the meaning clear. Rhetorical questions may be included to engage the reader. Information texts are usually written in third person and present tense. Dot points and sub-headings as well as paragraphs are acceptable. Your opening paragraph should provide a general overview followed by paragraphs, which provide detailed descriptions and a summary in conclusion.

Key terms: adverb, cohesion, context, exposition, informative, metaphor, narrative, point of view, pronoun, purpose, recount, rhetorical question, sequencing, simile, tense. See the Glossary on page 107 for definitions of these terms

Example Test: Annotated responses

Instructions

- Read the following annotated responses to see examples of each of the Informative, Narrative, Recount and Expository writing genres.

Narrative annotated example response

Visiting Pa and Gran

It was a warm Australian spring somewhere close to the Whitsunday Islands. Grandfather was sitting in his rocking chair half-dozing while watching **the sun rising** over his glistening ocean. The **happy** cries of children made him sit up straight. The hullabaloo was being caused by the arrival of **Trinh and George, his two grandchildren**. **Like waves on rocks** they dashed up the four steps onto the long verandah to kiss and hug him.

'Hello Pa!' **screamed** the two of them excitedly.

'How's my favourite gan'kids today?' **chuckled** their Pa.

'Fine, we are happy to see you!' they cheekily replied.

'Sorry we're late, Dad, we had a problem with the car,' **explained Jess**.

'Kids go say hello to Gran and I'll tell Pa all about it.'

Pa settled back in his chair gazing out over his personal paradise. The ebb and flow of the waves matched the rise and fall of his chest as he listened to their story.

Annotations:

- Introducing the setting and time of year.
- Characters introduced
- Establishes the time of day.
- Establishing tone.
- Uses simile to provoke image.
- Characters introduced
- Use of alternative word for 'said' to show emotion.
- Use of alternative word for 'said' to show reaction.
- When writing dialogue, each time the speaker changes, a new paragraph is required to indicate another character is now speaking.
- Use of alternative word for 'said' this is the complication in the structure of the text. This sentence flags the complication that is about to unfold.
- Characters introduced
- Sequences the text.
- Description adds to sense of grandfather being relaxed.

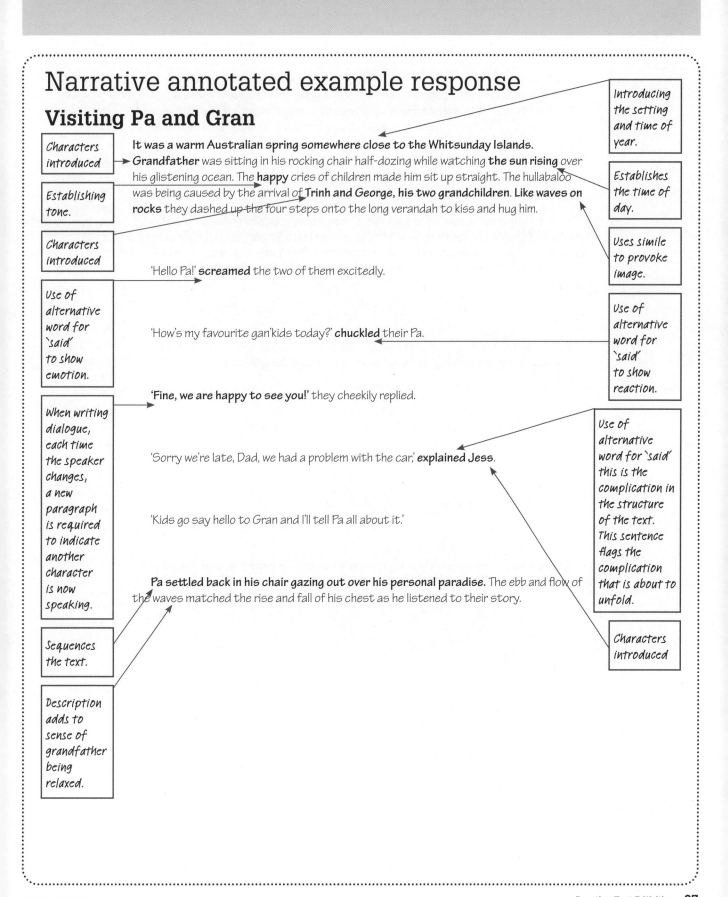

It was between **Brisbane and Noosa** and the traffic had crawled to a snail's pace. The children were impatient to reach their grandparents and were doing their usual squabbling over nothing in the back of the car. I changed the CD player back to the radio just in time to hear the **news . . .**

Our journey might have been over before we reached our destination. Hmmm, I wondered if that track through the pines we passed **awhile back** would still be drivable and we were in an all wheel drive vehicle. So I turned around and found the almost hidden track. It was more overgrown than I expected but we were still travelling quicker than the main road.

It was relaxing driving through the bush and the kids **became quiet and then excited**. We travelled through **lush, green ferns and monolithic trees**. It didn't look like anyone used this track anymore and the natural bush had grown almost across the road. The **wooshing** sound of the branches and ferns brushing along the sides of the car was lulling everyone into a state of almost total relaxation.

Wouldn't you know it; bang a flat tyre. A tree root had ripped open the side of the back tyre. We had to unload everything to get the spare and there was no room to move around because of the **dense undergrowth**. It took twice as long to change the wheel. No one wanted to help put everything back in the car so I had to do it all. **I had stopped feeling relaxed.**

Once we were driving north again the excitement returned and it was all good. **We found our way back onto the highway and here we are, a bit later than planned.**

Pa drifted back from his own thoughts, turned to Jess and asked 'What did you hear on the news, love?'

'Lunch is ready!' Gran shouted, tapping on the window.

Grandfather arose from his rocking chair and toddled towards the door, thinking he had a barrage of questions for the two children.

Building suspense through sequencing the events and leaving some events untold.

Defines the place for the complication, which is different to the setting of the grandparent's home.

Reference to the passing of time, sequencing the story

Maintaining cohesion.

Reinforces the sense of time passing.

Reinforces image of the surrounding area.

Personifies the sounds of the ferns and trees in the bush.

Highlights the tone change.

Resolving the issue of needing to take a detour. Conclusion to the trip, but does not complete the story.

Invites the reader to engage with the narrative and inferentially discuss what information had been heard on the car's radio.

9780170462198

Recount annotated example response

Diary Entry

Who and why – character and involvement.

What – the event that will be discussed.

Dear diary

I have tried to think of a good name for you but just can't!

Where – where it happened.

When – the timing of the event.

I can't believe **we lost our soccer match**, **in the hills**, **yesterday**. We should have won it, we were the best team. It was **all Ronaldo's fault**.

Sequencing word used to establish order.

Use of sarcastic tone.

Paints a clear picture of the thoughts and feelings of the players.

First he left all our club shirts at home. How were we supposed to play without our shirts!?! We tried to send him back for them but it was too late. So we ran onto the field, out of uniform. No one was happy we were **too embarrassed to be** happy. We were disorganised and unfocused.

Sequencing the game. Avoid using the word 'then' repeatedly to show the passing of time.

Provides another reason why the team lost the game.

The team expected everyone knew that we had to remember two things. Always stay in a useful position for the team and keep your eye on the ball. How many times had coach told us to play smart and anticipate? We had to limit our opponents' choices by reducing their space and covering their pass lines. **Did we do that? NO. Josh had a shocker he just could not get it together.**

Simile used to create an image of players who were stiff and not playing their best and fastest game in the reader's mind.

At half time coach gave us a pep talk but it didn't make any difference we played worse in the second half. Coach wanted us to cross the ball early from the flanks (wings) to the back of the defense! He reckoned that would work. It didn't. We **were like rusting robots and moved the same way.**

Use of the past tense is consistent and repeated to emphasise that the writer is recounting a previous event.

All that practice at training, we planned strategies and learned all the moves off by heart. It **was all Ronaldo's fault**. The crowd and our families laughed at our stupid outfits. We should have been wearing our uniforms, but we didn't have any. It was all Ronaldo's fault.

Colloquial language.

Another reason why the team lost the game.

Sarcastic tone used to engage the reader.

Gives an emotional response.

We knew that the other team would get confused if we changed places on the field. So we changed places. **Did it work? NO.** We confused ourselves and it was all Ronaldo's fault. Without our shirts how were we expected to know who was who. **It was just one big mess.**

Ronaldo refers to himself in the third person here.

Coach told us at half time try and put yourself inside the minds of your opponents and imagine what they were thinking. Big mistake. BIG mistake. It was all Ronaldo's fault he just stood there and laughed. He knew they thought we were the **lamest** soccer team they had ever played against. He didn't stop. That was the end of everything we lost the game because **no one played properly**.

Recognition to team mates that it was his poor team effort that caused them to lose the game.

Identifying Ronaldo as the diary writer. Change to first person shows acceptance of the writer's role in the team's loss.

I was so embarrassed. **It was all Ronaldo's** fault. I didn't **do my bit for the team** and we lost the game, made our whole club look stupid and I felt really bad. **I tried to say I was sorry** but nobody wanted to listen.

Bye, from Ronaldo.

Expository annotated example response

Anne of Green Gables (1985)

Produced and directed by Kevin Sullivan

States position for the rest of the film review clearly. Reader expectation is created.

Emotive language.

Highlights the importance of character and introduces the main character

It may not be the best ever film, but **I think I've just seen my favourite film** of the year, Anne of Green Gables. **This is a character-driven film about 11-year-old Anne Shirley**, an orphan, who is sent to live with an elderly brother and sister by mistake. However, she charms her new home and community with her **fiery spirit and imagination**. The film is closely based on Lucy Maud Montgomery's novel, published in 1908, and produced and directed by Kevin Sullivan in this 1985 film.

Narrative structure of the film. The complication. Note that this is not simply a plot summary.

Sixty year old Matthew Cuthbert, played by Richard Farnsworth, a soft spoken bachelor and his somewhat unhappy younger sister Marilla, played by Colleen Dewhurst, decide to adopt a young boy who can help out with chores around their farm. Matthew arrives at the train station expecting **to collect a sturdy boy instead he finds a skinny girl** with masses of red hair accompanied by little more than a worn out carpetbag and her imagination.

Descriptive and evocative introduction to the setting.

Present tense.

This 1985 version maintains the original sentiments of the novel. **The film is set** in the **picturesque rural village** of Avonlea on Prince Edward Island and Southern Ontario in Canada. The Green Gables house used in the Anne movies is actually two different buildings, both privately owned and both located in Southern Ontario. Much of the outdoors filming was completed before actors had been cast for the movie. Historic homes and villages around Ontario were used providing a sense of realism of Prince Edward Island in the 1900's.

Thought-provoking without giving away any specific details.

Narrative structure of the film, provides a resolution.

Good behaviour is a subject that troubles Anne but after each incident there is always a happy ending. Somehow Anne's irrepressible quirkiness manages to worm its way into the Cuthbert's hearts, and they soon prepare to **let her stay if she can manage to avoid trouble**. This proves to be a daunting task, as the headstrong quick-tempered girl has a remarkable knack for attracting calamity. Anne's defining characteristic is optimism in the face of uncertainty.

Indicates the genre.

Clue to content without spoiling or giving away specific details

Identifies the time period.

Expectation that viewers will feel empathy for the characters.

Links to historical context of community.

The emphasis on good manners and moral lessons may seem a bit oppressive in a modern society, but Anne's life is far from dull. It is a **feel good family movie** that invites **encouraging parallels between viewers and characters**. *Anne of Green Gables* encourages us to always look for the good in people and situations. It is a richly produced film which gives us a **sense of what life was like in rural Canada at the turn of the 20th century**.

Line of argument developed to encourage positive interest in the film and ending with an affirmative statement, which reinforces the point of view established in the introduction.

Despite a series of misadventures and theatrics arising from Anne's arrival in Avonlea this comical but enlightening story provides many inspirational moments. Filmed amidst the spectacular scenery of Prince Edward Island, Canada, this award-winning film follows the struggles of Anne through her **adolescence to her triumphs** as a young woman. **I highly recommend this film it is truly a classic full of wit, style and emotional** power.

9780170462198

Informative annotated example response

Life Cycle of a Butterfly

General information about butterflies – description – information is factual.

Butterflies are beautiful, flying insects with large scaly wings. They have six jointed legs, three body parts, a pair of antennae, compound eyes, and an exoskeleton. Butterflies can only fly if their body temperature is above 30 degrees. Butterflies sun themselves to warm up in cool weather. The life cycle of a butterfly is made up of four stages.

General information about butterflies – description – information is factual.

Additional information.

More detailed information on laying eggs.

Female butterflies lay eggs on the leaves of a host plant. A female butterfly can lay between 120 and several hundred eggs. The eggs appear to be small dots but they can vary in form, shell decoration, colour and size. Caterpillars grow inside the eggs and hatch after a few days.

Identifies the first stage.

Introduces the next stage.

Identifies the second stage.

Simple present tense.

Introduces the next stage.

Caterpillars hatch by eating their way out of their shell. They spend their whole life eating and growing. The first thing a caterpillar eats is its eggshell and then the leaves of its host plant. As the caterpillar grows it sheds its skins about five times. Caterpillars can eat approximately eight times their own body weight each day. The pupa is formed in the body of the caterpillar.

More detailed information on the birth of a caterpillar.

More information to explain what is happening at this stage.

As the caterpillar turns into the pupa it forms a hard shell called a chrysalis. This is the resting stage but there is a lot happening inside the chrysalis. Many changes happen to the chrysalis. The wings, legs and rest of the butterfly are formed inside the chrysalis.

Identifies the third stage.

More information on the life of a butterfly.

After 8–13 days of pupation, from the chrysalis emerges a beautiful adult butterfly. As an adult butterfly, it will eat nectar, which it finds by flying from flower to flower. Butterflies get the energy needed to reproduce and lay eggs for another generation.

Identifies the fourth stage.

Scientific information using technical language.

The life cycle or process of a caterpillar turning into a butterfly is called **metamorphosis**. Depending on the type of butterfly, the life cycle of a butterfly may take anywhere from one month to a year. There are approximately 28,000 species of butterflies around the world.

Conclusion and a short summary of the information provided in the text.

In the final stage of their life cycle the adults will reproduce and then die, beginning the life cycle again by laying eggs.

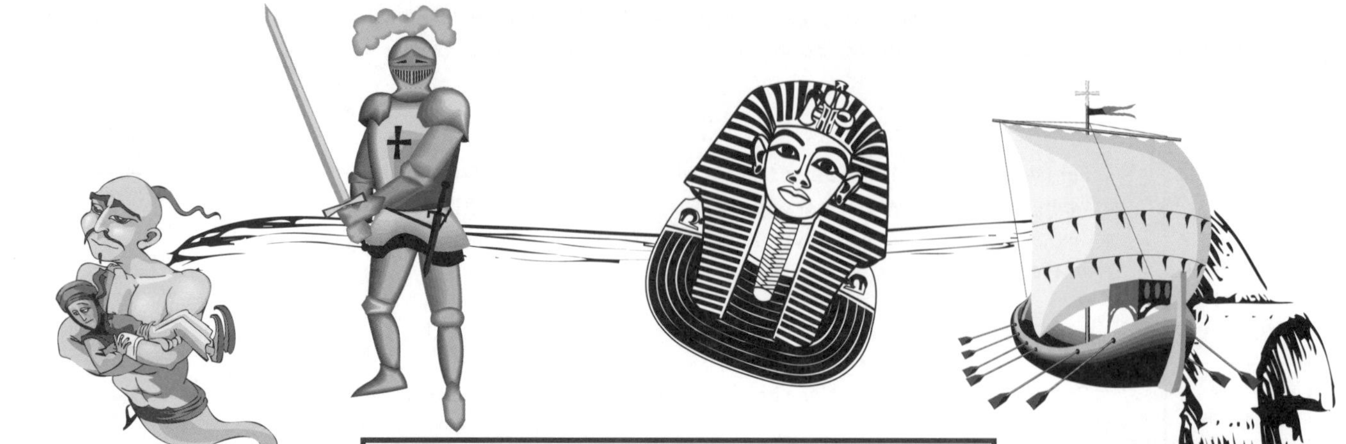

IMAGINE…

You are going to write a narrative or short story. The idea for your story is '**imagine**'. Imagine what might happen if you…

- found something amazing
- travelled to another land or planet
- travelled back in time
- travelled into the future
- invented something that could change the world
- found you had been granted three wishes

Planning your story:

- before you start, think about your characters, where they are, where they are going and how they will get there. What problems will they need to solve? Plan an exciting ending.

Crafting your story:

- use exciting and emotive words written in sentences and paragraphs
- use great action verbs
- spelling, punctuation and correct tense are all important and will help to create a more exciting story.

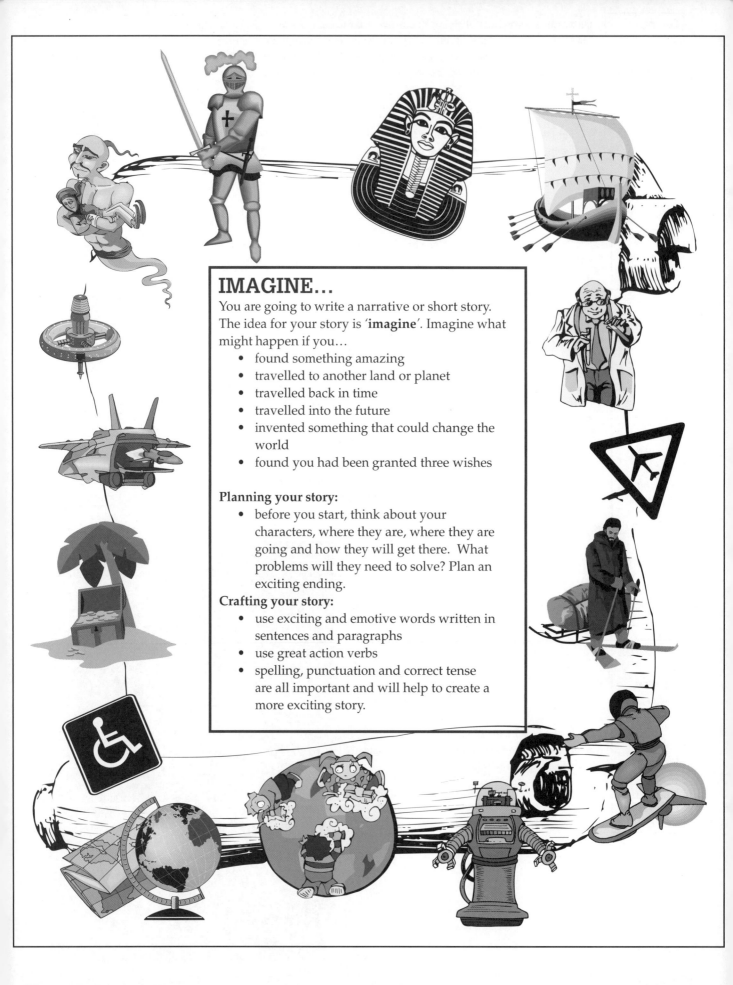

You are going to write a narrative or story.

The idea for your story is 'imagine'.

- Read through the ideas listed on the stimulus.
- Choose the ideas that appeal to you.
- You don't have to choose one of the ideas provided. They are offered to get you thinking quickly and assist in composing in a short, specific time frame.

Your own idea:

____

____

____

____

____

____

Planning your story:

- think about your characters. (Give your characters names, personalities and maybe a description.)
- where are they? (This is the setting.)
- include the main characters and/or setting in the introduction.
- where are they going and how will they get there? (This is part of the sequence of events that forms the body of your story.)
- what problems will they need to solve? (This is also part of the sequence of events that forms the body of your story.)
- think about your ending. How can you make it exciting, surprising and entertaining?

Crafting your story:

- use interesting and emotive words (include multi-syllable words and try not to be repetitive in your choice of words).
- use great action verbs (for example, think of other words to use instead of 'said' – such as 'whispered', 'stammered', 'screamed').
- spelling (be careful to spell words correctly).
- punctuation (capital letters, full stops, commas – all the required punctuation to assist the flow and cohesion of your narrative).
- correct tense (maintain the same tense throughout your story; decide whether it is past, present or future).

Writing your story:

- use a clear structure to build a cohesive and engaging narrative.
- use sentences that include a variety of simple, compound and complex and use correct punctuation.
- use paragraphs that begin with a topic sentence that introduces and develops a key thesis or central idea.

Checking your work:

Always read back over your writing to check for opportunities to gain extra marks by correcting your own mistakes, using a better word and different grammatical features.

- Spelling and punctuation – all correct?
- Vocabulary – could you use a different word to make your writing more interesting? Have you used the same word too many times?
- Using 'and', 'but', 'then' – have you done so too many times? Maybe you need to use a full stop and capital letter instead.
- Sentences – have you used a variety of simple, compound and complex?
- Paragraphs – insert a square bracket ([) to indicate a new paragraph without rewriting.
- Tense – is it consistent throughout your story?

Criteria:

There are ten criteria assessed in the writing task:

- audience
- text structure
- characters
- events
- vocabulary
- sentence structure
- paragraphs
- cohesion
- punctuation
- spelling.

Sample Writing Test – Narrative

Look at the stimulus. Brainstorm your ideas and then use this box to write a clear plan. Allow no more than five minutes. This planning will not be marked.

9780170462198

9780170462198

Year 7 Literacy

Language Conventions Test 1

Writing time: 45 minutes

Use 2B pencil only

Instructions

- Write your **student name** in the space provided.
- You must be silent during the test.
- If you need to speak to the teacher, raise your hand. Do not speak to other students.
- Answer all questions using a 2B pencil.
- If you wish to change your answer, erase it very thoroughly and then write your new answer.

Student name:

Each sentence has one word that is incorrect.
Write the correct spelling of the word in the box.

WRITE YOUR OWN ANSWER

1 I chose to study Sience at school.

2 There could be no doubt wich flavour I preferred.

3 Many woman catch the bus to town.

4 He acknowleged each guest with a tip of his hat.

5 I brought a magazine with money I had saved.

6 I went to the doctor to complane about my cough.

7 The monster in the scary film horified me.

Each sentence has one word that is incorrect.
Write the correct spelling of the word in the box.

WRITE YOUR OWN ANSWER

8 I had a great idear for the school play.

9 I enjoy going to the libary to use the Internet and choose a book to read.

10 My shoelace was quite lose, so I stopped to tie it.

11 I like orange juise with my breakfast.

12 The explotion of fireworks lit up the sky.

13 The textbook used a clear exsample of an experiment for the student to follow.

14 My friend had forgoten his lunch.

The spelling mistakes in these texts have been circled.
Write the correct spelling for each circled word in the box.

WRITE YOUR OWN ANSWER

15 The pyramid was jigantic,

goldin in colour, and its

enormos shadow stretched out

upon the sandy dessert.

9780170462198

WRITE YOUR OWN ANSWER

16 The leafes on the newly planted

potatos had wilted as a result of

the absense of rain during

the most recent drowt.

WRITE YOUR OWN ANSWER

17 The resent law passed by the local

goverment ensured that neighbours

would have to be responsable for

maintaining the boundary fences.

18 Which of the following completes the sentence?

SHADE ONE BOX

Baz Luhrmann's epic film *Australia*, ☐ was filmed in Darwin, featured prominent landmarks of Northern and Western Australia.

☐ what ☐ who ☐ that ☐ where

19 Which of the following correctly completes the sentence?

SHADE ONE BOX

She wanted to hear who had won the competition, so she ☐ turned on the radio.

☐ quick ☐ quicker ☐ quickest ☐ quickly

20 Which of the following correctly completes the sentence?

SHADE ONE BOX

Spring flowers are ☐ than flowers blooming at any other time of year.

☐ prettiest ☐ most prettiest ☐ pretty ☐ prettier

21 Which of the following correctly completes the sentence?

SHADE ONE BOX

He was [ ] his shirts daily for work.

☐ ironing ☐ ironed ☐ irons ☐ had ironed

22 Which of the following correctly completes the sentence?

SHADE ONE BOX

I had [ ] enough biscuits yesterday to feed the whole group.

☐ baking ☐ bake ☐ baked ☐ bakes

23 Which sentence is correct?

SHADE ONE BOX

☐ The annual flooding of the desert weren't necessary for the flora and fauna to survive.

☐ The annual flooding of the desert are necessary for the flora and fauna to survive.

☐ The annual flooding of the desert were necessary for the flora and fauna to survive.

☐ The annual flooding of the desert is necessary for the flora and fauna to survive.

24 Which sentence is correct?

SHADE ONE BOX

☐ The argument is between you and it.

☐ The argument is between you and me.

☐ The argument is between I and you.

☐ The argument is between you and myself.

25 Last night, to help me relax, I [ ] television before going to bed.

SHADE ONE BOX

☐ watched ☐ watch ☐ watches ☐ watching

26 Which sentence has the correct punctuation?

SHADE ONE BOX

☐ The Sydney Opera House is to be closed for refurbishment in January

☐ The Sydney Opera House is to be closed for refurbishment in January.

☐ The sydney Opera House is to be closed for refurbishment in January.

☐ The Sydney Opera House is to be closed for refurbishment in january.

9780170462198

27 Which of the following correctly completes the sentence?

SHADE ONE BOX

After his bull had won the Blue Ribbon, he was the [ ] boy at the town fair.

- [ ] happy
- [ ] happiest
- [ ] happier
- [ ] most happiest

28 Which of the following correctly completes the sentence?

SHADE ONE BOX

We [ ] prepared for the destructive storm, because it blew in so quickly.

- [ ] were
- [ ] weren't
- [ ] were'nt
- [ ] where

29 Which of the following correctly completes the sentence?

SHADE ONE BOX

Though it had a small courtyard, it was the [ ] building I had ever seen.

- [ ] kindest
- [ ] tallest
- [ ] happiest
- [ ] longest

30 Which of the following correctly completes the sentence?

SHADE ONE BOX

The plane above us [ ] quickly into the atmosphere.

- [ ] climbed
- [ ] climb
- [ ] climber
- [ ] climbing

31 Which sentence has the correct punctuation?

SHADE ONE BOX

- [ ] He said, 'to return the last book to the shelf before the library closes.'
- [ ] He said to return the last book to the shelf before the library closes.
- [ ] He said to, 'return the last book to the shelf before the library closes.'
- [ ] He said to, 'return the last book to the shelf,' before the library closes.

32 Which of the following correctly completes the sentence?

SHADE ONE BOX

At sunset, the flock of birds screeched [ ] before they settled for the night.

- [ ] loud
- [ ] louder
- [ ] loudly
- [ ] loudest

33 Which of the following correctly completes the sentence?

SHADE ONE BOX

The television's picture [ ] in the last year to the point that images were only in black and white.

☐ has worsened ☐ worser ☐ had worsened ☐ worst

34 Which of the following correctly completes the sentence?

SHADE ONE BOX

The star of the film [ ] buying groceries at the local store last week.

☐ is seen ☐ seeing ☐ was saw ☐ was seen

35 Which sentence has the correct punctuation?

SHADE ONE BOX

☐ 'I'll run to second base I said.'
☐ 'I'll run,' to second base I said.
☐ I'll run to second base I said.
☐ 'I'll run to second base,' I said.

36 Which sentence has the correct punctuation?

SHADE ONE BOX

☐ They ran to the shop, before the movie started to buy some popcorn.
☐ They ran to the shop before the movie started, to buy some popcorn.
☐ They ran, to the shop before the movie started to buy some, popcorn.
☐ They ran to the shop, before the movie started, to buy some popcorn.

37 Shade one box to show where the missing apostrophe (') should go.

SHADE ONE BOX

A family☐s pet☐s: their cat☐s, dogs and bird were named after famous movie stars☐.

38 Which sentence has the correct punctuation?

SHADE ONE BOX

☐ The boy shouted, 'Turn around and you'll see the helicopter!'
☐ The boy 'Shouted turn around and you'll see the helicopter!'
☐ The boy shouted turn around and you'll see the helicopter!
☐ The boy shouted, 'Turn around,' and you'll see the helicopter!

39 Shade one box to show where the missing apostrophe (') should go.

SHADE ONE BOX

The car☐s passenger☐s asked the driver to change the music as the station☐s that they had been tuning into played old-fashioned song☐s.

9780170462198

40 Which sentence has the correct punctuation?

SHADE ONE BOX

- ☐ The girl who had just completed the marathon, was offered a bottle of water and a place to sit.
- ☐ The girl, who had just completed the marathon, was offered a bottle of water and a place to sit.
- ☐ The girl who had just completed the marathon was offered, a bottle of water, and a place to sit.
- ☐ The girl, who had just completed the marathon was offered a bottle of water, and a place to sit.

41 Which sentence has the correct punctuation?

SHADE ONE BOX

- ☐ Kate said that we had, 'better wash our hands before dinner.'
- ☐ Kate said that, 'we had better wash our hands before dinner.'
- ☐ Kate said that we had better wash our hands before dinner.
- ☐ Kate said, 'that we had better wash our hands before dinner.'

42 Shade **two** boxes to show where the missing commas (,) should go.

SHADE TWO BOXES

A number ☐ of environmental studies ☐ which have been carried out over recent years ☐ indicate that ☐ the use of plastic bags has decreased.

Year 7 Literacy

Reading Magazine 1

Backpacks – or how to carry your life around with you

However did we manage before they were invented!

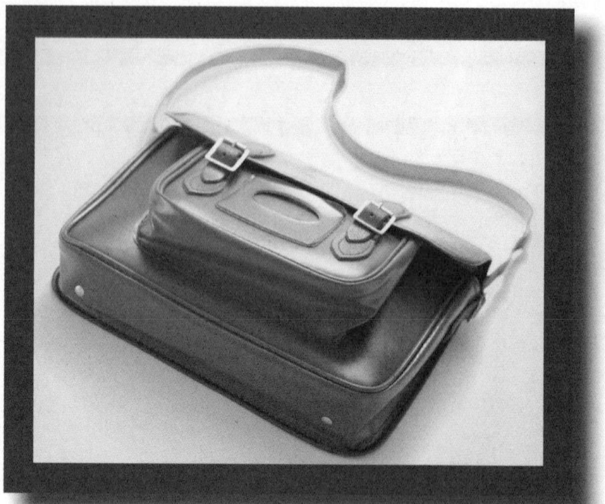

The old days of school satchels – a sort of leather bag with a long handle that hung off the shoulder – must have led to heaps of kids walking lopsidedly. Actually, what used to happen was that kids sort of hugged the satchel or balanced it on one hip, which meant that hands were busy holding onto the bag and hips got sore!

When backpacks can be dangerous

Yes, backpacks are great, but you do need to take a bit of care or they could be dangerous.

- It could fall on someone if it is on top of your locker or a shelf at school.
- The straps could trip someone if they are sticking out from the bag.
- Not wearing them properly can hurt your back, neck or shoulders.

What else you need to know about backpacks

Backpacks with just one big pocket mean that everything gets lumped in together.

Choose one with several pockets to organise your stuff better. Then the weight will be spread out more evenly, which will be better for you to carry.

If you do have a lot to carry and there are no lockers at your school, then it may be a good idea to get a backpack that has wheels and a pull out handle.

And then of course, there are all those pockets in a backpack, so that you can organise things and find them easily.

Dr Kate says:

'The most you should carry is around 10% to 15% of your body weight – so if you're a little kid and you weigh 32kg, you should only be carrying between 3 and 4kg of weight. Carrying more can be doing damage to your back.

Tell a parent if you are getting headaches, backaches or tingly feelings in your arms or back – you may not have adjusted the straps properly or you could be carrying too much weight.'

9780170462198

Ice Cream

Page 1 of 2

Ice Cream
It's in the bag

Make your favourite ice cream without going near a freezer!

Educators' notes at www.sciencemuseum.org.uk/educatorsresources

sciencemuseumlearning

GRAB THIS STUFF...

- Half a cup of milk (plain or flavoured)
- Ice
- 1 tablespoon of caster sugar
- 6 tablespoons of salt
- Small zip-seal bag
- Large zip-seal bag
- Half a teaspoon of vanilla essence (optional)

Icing Sugar
TABLE SALT

Grab a spoon and enjoy your ice cream!

Page 2 of 2

Ice Cream
It's in the bag

Make your favourite ice cream without going near a freezer!

Educators' notes at www.sciencemuseum.org.uk/educatorsresources

sciencemuseumlearning

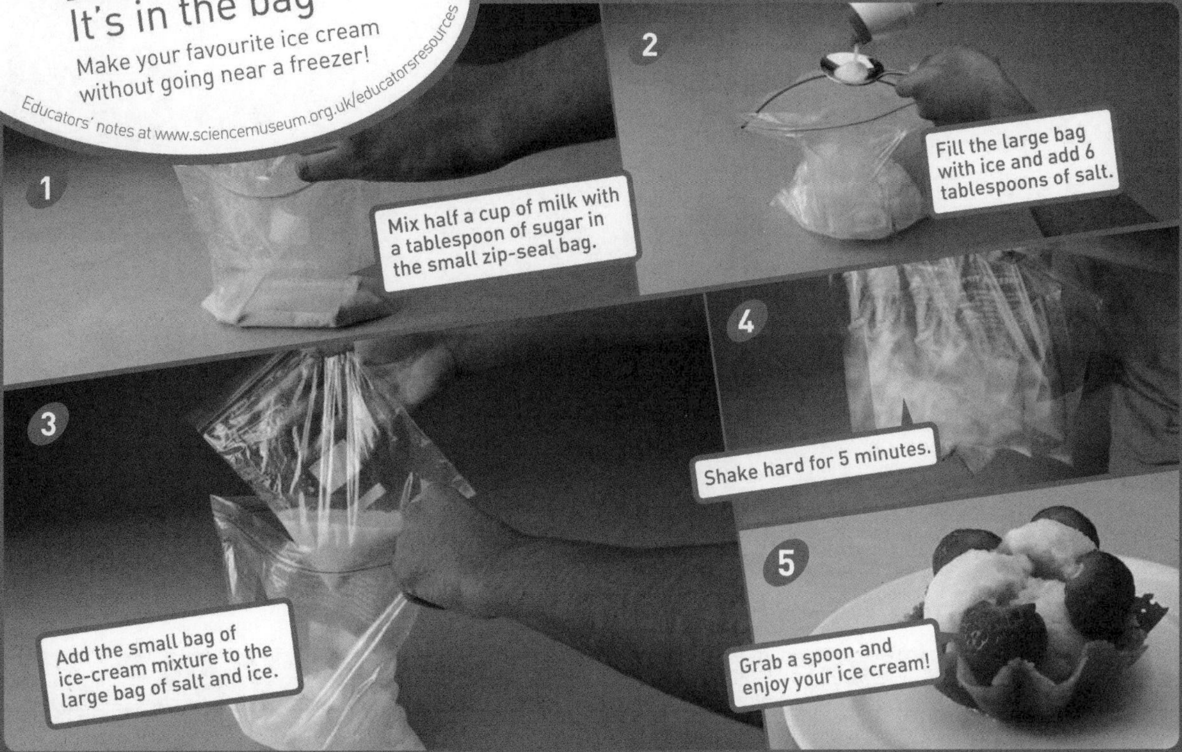
1
Mix half a cup of milk with a tablespoon of sugar in the small zip-seal bag.
2
Fill the large bag with ice and add 6 tablespoons of salt.
3
Add the small bag of ice-cream mixture to the large bag of salt and ice.
4
Shake hard for 5 minutes.
5
Grab a spoon and enjoy your ice cream!

Albert Einstein

Albert Einstein was a German-born physicist, although most people probably know him as the most intelligent person who ever lived. His name has become part of many languages when we want to say someone is a genius, as in the phrase, 'She's a real Einstein'. He must have been pretty brainy to discover the Theory of Relativity and the equation $E=mc^2$.

In 1999, 'Time' magazine named Einstein as the Person of the Century. No one could have guessed this would happen when he was at school. He was extremely interested in science but hated the system of learning by heart. He said it destroyed learning and creativity. He had already done many experiments, but failed the entrance exams to a technical college.

He didn't let this setback stop him. When he was 16, he performed his famous experiment of imagining travelling alongside a beam of light. He eventually graduated from university, in 1900, with a degree in physics. Twelve years later he was a university professor and in 1921, he won the Nobel Prize for Physics. He went on to publish over 300 scientific papers.

Einstein is the only scientist to become a household name, and part of everyday culture. He once joked that when people stopped him in the street, he always replied: 'Pardon me, sorry! Always I am mistaken for Professor Einstein.' Today, he is seen as the typical mad, absent-minded professor, who just happened to change our world.

9780170462198

Robots

FILM SYNOPSIS

Twentieth Century Fox, Blue Sky Studios and Academy Award®-winning director Chris Wedge, who transported audiences to prehistoric times with their box-office smash Ice Age™, have now created the visually spectacular world of Robots™ a world filled entirely with whimsical robots. Like Ice Age™, the movie is packed with comedy, incredible visuals and a lot of heart.

Rodney Copperbottom™, voiced by Ewan McGregor, is a small town robot who has a gift for inventing things, but is trapped in the confines of his quaint surroundings. He works side by side in a restaurant with his dad who is a dishwasher - literally a dishwasher. You open his chest and load in the dishes. Rodney has dreams of something greater.

Armed with his unique talent for inventing, Rodney embarks on a journey to Robot City to meet his idol, the majestic inventor Bigweld, voiced by Mel Brooks. An iconic figure in all of Robot City, Bigweld has spent a lifetime creating things to make the lives of robots better. Once in Robot City, Rodney finds that things are not quite as he expected, and his quest may be a lot harder than he imagined.

As he tries to navigate his way around this new city, Rodney befriends the Rusties, a ragtag group of street-smart bots who know the ropes. One of the Rusties, Fender (voiced by Robin Williams), immediately becomes Rodney's best friend and even lets his kid sister Piper (voiced by Amanda Bynes) tag along. They take him in, and for now, at least, Rodney has a home in Robot City.

Rodney also meets Cappy (voiced by Halle Berry), an executive at Big Weld Industries who takes an instant liking to Rodney and sees a lot of herself in him. Along their adventures, Rodney and his new friends encounter unsavory characters who try to derail Rodney's plans to find Bigweld and save Robot City. The result is a timeless, comedic tale that pushes the boundaries of animation while introducing characters rich with humor and soul, and a heart-warming story that proves that a robot can shine no matter what he is made of.

Lord Howe Island

Permanent Park Preserve

Lord Howe Island is an outstanding natural area of national and international significance. This significance is recognised by the island's status as a property on the World Heritage List.

The Lord Howe Island Permanent Park Preserve (PPP) was dedicated to protect the unique natural values of Lord Howe Island and neighbouring Islands. The PPP covers 75% of Lord Howe Island, including the southern mountains and northern hills. The PPP also includes Balls Pyramid and neighbouring Islands.

The PPP is similar to a National Park in terms of the primary management emphasis is directed at conservation and preservation of natural values, the main difference being the PPP is managed by the Lord Howe Island Board rather than the NSW National Parks & Wildlife Service.

The PPP is of outstanding value for nature conservation, for aesthetic appreciation, for recreation, for education and for research.

There are over 200 native species of vascular plants on the Island. Over 70 plants are endemic to the Island (meaning Lord Howe Island is the only place in the world where they grow naturally). NPWS are currently undertaking a comprehensive endemic plant survey to determine if some of these plants should be nominated as threatened under the NSW Threatened Species Conservation Act 1995. Lord Howe has 129 plant genera in common with Australia, 102 with New Caledonia and only 75 in common with New Zealand. Due to the high level of endemism the Island flora is of outstanding regional nature conservation value.

9780170462198

Macbeth and Son

Jackie French

Luke shoved his copy of *Macbeth* to the back of his desk. What did all those words mean, anyway? he thought as he opened the window. His bedroom stank of air freshener. It always had that not-quite-roses smell after Mrs Tomlin cleaned.

Mrs Tomlin and her husband lived in the cottage down past the machinery shed. The cottage had just been a wreck when Mum and Dad had the farm. But when Sam married Mum he'd had the cottage renovated at the same time as the new wing of the house was built.

Now Mrs Tomlin did the housework, and the cooking too when Mum went down to stay with Sam during the week in Sydney, and Mr Tomlin helped Mum run the farm. Mum and Sam slept in the new part of the house, but Luke had kept his old bedroom.

He breathed in the night air gratefully. Cold cowpat wasn't the best smell in the world. But at least it was a real smell. Better than air freshener.

It was three days now since the letter had come from St Ilf's. Three days of people congratulating him, telling him 'Well done.' Three days of empty triumph. Mum had been walking around with a grin the entire time, singing 'Rocky Mountain High' under her breath. You always knew Mum was over the moon when she sang 'Rocky Mountain High.'

How could he ever tell Mum he'd cheated?

But he hadn't really cheated, he told himself. Cheating was when you meant to do it. How was he to know that the exam paper would be one he'd seen before?

If only he hadn't won the scholarship! If he'd just passed the entrance exam it wouldn't have been so bad. It wasn't even as though he needed it. Sam had plenty of money. And now everyone would expect this brilliant kid and instead they'd just get dumb old Luke. He didn't want to go to school in Sydney, away from all his friends.

If only he'd mentioned that he'd already seen the exam paper immediately . . . or at least when the letter came. But if he said anything now everyone would think he *had* cheated. And it'd break Mum's heart . . .

. . . What would he do?

At least in Shakespeare's world things were clear, thought Luke, almost asleep. You knew what was right and wrong in those days . . . If only I lived in a world like that . . .

New Year's Watch

Su Shi

Soon now, we'll mark the year's end that approaches.
It's like a snake that crawls into a hole.
Already half its scaly length is hidden
What man can stop us losing the last trace?
And even if we wanted to tie its tail
No matter how we try, we can't succeed.
The children make all effort not to sleep
We laugh together, watching though the night.
The cockerels should not cry the dawn for now
The drums as well should give the hour
respect.
We sat so long the lamp's burnt down to ash
I rise and see the Plough is slanting north.
Next year, perhaps, my span of years could
end
My fear is that I've just been marking time.
So exert ourselves to the utmost here tonight
I still admire the exuberance of our youth!

Hamilton, Elaine and Farr, Robin. Poetry Unlocked, Farr Books, Queensland, 2008.

Looking for Alibrandi

Melina Marchetta

My name is Josephine Alibrandi and I turned seventeen a few months ago . . .

We live in Glebe, a suburb just outside the city centre of Sydney and ten minutes away from the harbour. Glebe has two facades.

One is of beautiful tree-lined streets with gorgeous old homes and the other, which is supposed to be trendy, has old terraces with views of out-houses and clothes-lines. I belong to the latter. Our house is an old terrace. We, my mother Christina and I, live on the top. We were actually renting the place until I was twelve but the owner sold it to us for a great price and although I've calculated that Mama will have it paid off when I'm thirty-two, it's good not to be renting in these days of housing problems.

My mother and I have a pretty good relationship, if a bit erratic. One minute we love each other to bits and spend hours in deep and meaningful conversation and next minute we'll be screeching at each other about the most ridiculous things, from my room being in a state of chaos to the fact that she won't let me stay overnight at a friend's home.

She works as a secretary and translator for a few doctors in Leichhardt, a suburb unfortunately close to my grandmother's home, which means I have to [go] straight to Nonna's in the afternoon and wait for her. That really gets on my nerves. Firstly, the best-looking guys in the world take the bus to Glebe while the worst take the bus to where my grandmother lives. Secondly, if I go straight home in the afternoon I can play music full volume whereas if I go to Nonna's the only music she has is *Mario Lanza's Greatest Hits.*

My mother is pretty strict with me. My grandmother tries to put her two cents worth in as well, but Mama hates her butting in. The two of them are forever at loggerheads with each other.

Year 7 Literacy

Reading Test 1

Writing time: 65 minutes

Use 2B pencil only

Instructions

- Write your **student name** in the space provided.
- You must be silent during the test.
- If you need to speak to the teacher, raise your hand. Do not speak to other students.
- Answer all questions using a 2B pencil.
- If you wish to change your answer, erase it very thoroughly and then complete your new answer.

Student name:

Text 1: *Backpacks* Questions

Shade one box to show the correct answer to the following questions.

1 The main message in this text is that

SHADE ONE BOX

☐ backpacks are too dangerous for young children.

☐ backpacks are more popular than satchels.

☐ backpacks are hard to keep organised.

☐ backpacks must be chosen carefully.

2 According to the text backpacks can be dangerous if

SHADE ONE BOX

☐ they have too many pockets.

☐ they get stuck in your locker.

☐ they are balanced on your hip.

☐ they cause someone to fall over.

3 Backpacks often have a lot of pockets so that

SHADE ONE BOX

☐ you can carry more things to school.

☐ you can easily find your things.

☐ the weight will not spread out.

☐ teachers can give more homework.

4 If you weigh 32 kilos then your backpack should not weigh more than

SHADE ONE BOX

☐ 1 kilo ☐ 4 kilos ☐ 10 kilos ☐ 15 kilos

5 According to the text, a tingly feeling in your arms or back could be a sign that the straps

SHADE ONE BOX

☐ are too wide.

☐ should be flipped.

☐ need to be changed.

☐ are not strong enough.

9780170462198

Text 2: *Ice Cream* Questions

Shade one box to show the correct answer to the following questions.

1 The *Ice Cream* text is

SHADE ONE BOX

- ☐ a narrative.
- ☐ a procedure.
- ☐ a recount.
- ☐ an argument.

2 The text says, 'Ice cream. It's in the bag.' This is to

SHADE ONE BOX

- ☐ show where you will find the ice cream.
- ☐ describe how best to store the ice cream.
- ☐ make a play on words about the recipe.
- ☐ highlight that two bags are needed in the recipe.

3 One ingredient in the ice cream recipe is different in the list to its picture. Which ingredient is different?

SHADE ONE BOX

- ☐ the milk
- ☐ the sugar
- ☐ the salt
- ☐ the butter

4 How many ingredients are needed to make this ice cream?

☐ three ☐ four ☐ eight ☐ ten

SHADE ONE BOX

5 This recipe wants you to learn about

SHADE ONE BOX

- ☐ healthy eating.
- ☐ how things work.
- ☐ saving electricity.
- ☐ the reuse of materials.

6 'Half a teaspoon of vanilla essence (optional)'.

'Optional' means

SHADE ONE BOX

- ☐ unusual.
- ☐ necessary.
- ☐ not important.
- ☐ needs to be carefully measured.

Text 3: *Albert Einstein* Questions

Shade one box to show the correct answer to the following questions.

1 Where was Albert Einstein born?

SHADE ONE BOX

- ☐ Austria
- ☐ Germany
- ☐ Australia
- ☐ America

2 According to the text Einstein learned best through

SHADE ONE BOX

- ☐ repetition.
- ☐ creativity.
- ☐ reading books.
- ☐ memory.

3 Einstein won the Nobel Prize in

SHADE ONE BOX

- ☐ 1900.
- ☐ 1912.
- ☐ 1921.
- ☐ 1999.

4 Do you think Einstein had a sense of humour?
Use evidence from the text to support your answer.

WRITE YOUR OWN ANSWER

____

____

____

5 Albert Einstein will be remembered as someone

SHADE ONE BOX

- ☐ who was a comedian.
- ☐ who changed humanity.
- ☐ who was always absent-minded.
- ☐ who failed all his exams.

6 According to the text 'a real Einstein' is

SHADE ONE BOX

- ☐ an absent minded-professor.
- ☐ someone who hates exams.
- ☐ a very clever person.
- ☐ someone who reads a lot.

9780170462198

Text 4: *Robots* Questions

Shade one box to show the correct answer to the following questions.

1 What is a film synopsis?

SHADE ONE BOX

- ☐ The blurb for the cover of the DVD pack.
- ☐ The summary of a story told in present tense.
- ☐ The first draft of a story to be made into a film.
- ☐ The credits shown at the beginning of a movie.

2 The purpose of this text is to

SHADE ONE BOX

- ☐ persuade.
- ☐ preview.
- ☐ recount.
- ☐ record.

3 According to the text, 'created the visually spectacular world of *Robots* a world filled entirely with whimsical robots' describes

SHADE ONE BOX

- ☐ a world that is suddenly made by robots.
- ☐ a boring film about how robots live.
- ☐ robots taking over the world.
- ☐ an excellent film that encourages your imagination.

4 Rodney Copperbottom is

SHADE ONE BOX

- ☐ a robot who invents things.
- ☐ the director of the film.
- ☐ a famous Hollywood actor.
- ☐ a robot trapped in a small town.

5 Rodney sets out on a quest to meet

SHADE ONE BOX

- ☐ Mel Brooks.
- ☐ Cappy.
- ☐ the Rusties.
- ☐ Bigweld.

6 Who becomes Rodney's best friend?

SHADE ONE BOX

- ☐ Fender
- ☐ Cappy
- ☐ Bigweld
- ☐ Piper

7 'The result is a timeless, comedic tale that pushes the boundaries of animation while introducing characters rich with humour and soul, and a heart-warming story that proves that a robot can shine no matter what he is made of.' The purpose of these comments is to persuade you to see the film. Rewrite this statement in your own words.

WRITE YOUR OWN ANSWER

____

____

____

____

Text 5: *Lord Howe Island* Questions

Shade one box to show the correct answer to the following questions.

1 According to the text, Lord Howe Island's importance is shown by it being

SHADE ONE BOX

- ☐ a great place to go for a holiday.
- ☐ a regional conservation park.
- ☐ on the World Heritage List.
- ☐ on the NSW Property List.

2 Permanent Park Preserve is an example of

SHADE ONE BOX

- ☐ caring for the environment for the future.
- ☐ explaining why kids need parks.
- ☐ why we should put rubbish in the bin.
- ☐ advertising a holiday destination.

3 Lord Howe Island has more plants in common with

SHADE ONE BOX

- ☐ Australia.
- ☐ New Zealand.
- ☐ New Caledonia.

4 In this text the word 'endemic' means

SHADE ONE BOX

- ☐ that these plants could make you very sick if you eat them.
- ☐ this is the only place in the world where they grow naturally.
- ☐ that there are more than 200 different species on the island.
- ☐ plants that will only grow on the northern part of the island.

5 The PPP covers

SHADE ONE BOX

- ☐ three-quarters of Lord Howe Island.
- ☐ protection of 200 different plants.
- ☐ the neighbouring islands.
- ☐ all the mountains on the island.

9780170462198

6 According to the text the main difference between a National Park and a Permanent Park Preserve is

SHADE ONE BOX

- ☐ the value to the community.
- ☐ the size of the area each covers.
- ☐ who has the overall responsibility.
- ☐ the number of threatened species.

7 According to the text, much of the island flora is of particular conservation interest because the flora

SHADE ONE BOX

- ☐ is not found elsewhere.
- ☐ is very brightly coloured.
- ☐ only grows in mountainous areas.
- ☐ grows much faster than standard plants.

Text 6: *Macbeth and Son* Questions

Shade one box to show the correct answer to the following questions.

1 Luke shoved his copy of Macbeth to the back of his desk because

SHADE ONE BOX

- ☐ he did not like studying Shakespeare.
- ☐ he wanted to take good care of the play.
- ☐ he was feeling guilty.
- ☐ he wanted to go and play with friends.

2 'Three days' is repeated in the extract to

SHADE ONE BOX

- ☐ emphasise a specific time period in the story.
- ☐ make more lines to fill up space in a paragraph.
- ☐ help Luke remember how long something had been.
- ☐ show Luke missed his mother when she was away.

3 Luke's Mum sang 'Rocky Mountain High' when she was

SHADE ONE BOX

- ☐ angry.
- ☐ happy.
- ☐ confused.
- ☐ relaxed.

4 Luke thinks it's not cheating if

SHADE ONE BOX

- ☐ you didn't mean to do it.
- ☐ you don't win a scholarship.
- ☐ you don't tell anyone.
- ☐ no one finds out.

5 Luke didn't need the scholarship because

SHADE ONE BOX

- [ ] he hated school and was leaving.
- [ ] he already had one.
- [ ] his Mum needed help on the farm.
- [ ] he could afford the fees.

6 The purpose of this text is to show readers that

SHADE ONE BOX

- [ ] studying Shakespeare's *Macbeth* is no fun.
- [ ] new additions to families are hard to adjust to.
- [ ] cheating causes lots of problems for everyone involved.
- [ ] living on a farm is much better than living in the city.

7 Luke did not understand Shakespeare. True or False? Use evidence from the text to support your answer.

____

____

____

____

Text 7: *New Year's Watch* Questions

Shade one box to show the correct answer to the following questions.

1 This poem is about

SHADE ONE BOX

- [ ] the end of another year.
- [ ] an animal.
- [ ] a death in the family.
- [ ] having a party.

2 'It's like a snake that crawls into a hole' is a simile used to show

SHADE ONE BOX

- [ ] you might see a snake at night.
- [ ] it's dark on New Year's Eve.
- [ ] you can't see very well at night.
- [ ] time keeps on moving forward.

3 Explain in your own words the meaning of these three lines:

What man can stop us losing the last trace?
And even if we wanted to tie its tail
No matter how we try, we can't succeed.

WRITE YOUR OWN ANSWER

Write the answer on the lines.

____

____

____

____

4 A cockerel is a

SHADE ONE BOX

- ☐ rooster.
- ☐ chicken.
- ☐ duck.
- ☐ rabbit.

5 The phrase 'my span of years' tells us the person is

SHADE ONE BOX

- ☐ a young child.
- ☐ someone older.
- ☐ a local musician.
- ☐ a female dancer.

6 The purpose of this poem is to

SHADE ONE BOX

- ☐ encourage everyone to enjoy every day.
- ☐ celebrate at a New Year's Eve party.
- ☐ encourage people to care about their families.
- ☐ explain why the cockerel wouldn't call.

Text 8: *Looking for Alibrandi*
Questions

Shade one box to show the correct answer to the following questions.

1 When did Josephine turn 17?

SHADE ONE BOX

- ☐ last month
- ☐ she is about to turn 17
- ☐ a few months ago
- ☐ last year

2 In the text 'facade' means

SHADE ONE BOX

- ☐ the outsides of houses in Glebe are very beautiful.
- ☐ the fronts of the houses are beautiful but their backyards might be ugly.
- ☐ the two-storey houses allow you to see Sydney Harbour from the top floor.
- ☐ the outsides of houses all look the same.

3 Josephine and her mother

SHADE ONE BOX

- [ ] are always arguing.
- [ ] always agree with each other.
- [ ] don't like each other.
- [ ] get on well but sometimes argue.

4 Josephine wishes she could take the bus home after school because

SHADE ONE BOX

- [ ] she has a lot of homework she needs to get finished.
- [ ] it takes much longer to get home after school.
- [ ] she is seventeen and wants to get a part-time job
- [ ] the boys who catch the bus to Glebe are the best looking.

5 How well do Josephine's mother and her grandmother get on? Use evidence from the text to support your answer.

____

____

____

____

9780170462198

Year 7 Literacy

Writing Test 1

Writing time: 40 minutes

Use 2B pencil, blue or black pen only

Instructions

- Write your **student name** in the space provided.
- You must be silent during the test.
- If you need to speak to the teacher, raise your hand. Do not speak to other students.
- Use a 2B pencil or a black or blue pen only.
- Use the lines provided. Do not write in the borders.

Student name:

You are going to write a personal recount.

The idea for your recount is 'my holiday'.

- Read through the prompts listed on the stimulus.
- Use the prompts to plan your story.
- You don't have to use all of the prompts provided. They are offered to get you thinking quickly and assist in composing in a short, specific time frame.

Your own idea:

____

____

____

____

____

____

Planning your recount:

- think about your characters, the people you went with or met on holiday. (Give your characters names, personalities and maybe a description.)
- where did you go? (This is the setting.)
- make sure that the main characters and/or setting are included in the introduction.
- where are they going and how will they get there? (This is part of the sequence of events that forms the body of your story.)
- what events or problems did you encounter? (This is also part of the sequence of events that forms the body of your story.)
- think about your ending. How can you make it exciting, surprising and entertaining?

Crafting your recount:

- use interesting and emotive words (include multi-syllable words and try not to be repetitive in your choice of words).
- use great action verbs (for example, think of other words to use instead of 'said' – such as 'whispered', 'stammered', 'screamed').
- spelling (be careful to spell words correctly).
- punctuation (capital letters, full stops, commas – all the required punctuation to assist the flow and cohesion of your narrative).
- correct tense (maintain the same tense throughout your story; decide whether it is past, present or future).

Writing your recount:

- use a clear structure to build a cohesive and engaging recount.
- use sentences that include a variety of simple, compound and complex and use correct punctuation.
- use paragraphs that begin with a topic sentence that introduces and develops a key thesis or central idea.

Checking your work:

Always read back over your writing to check for opportunities to gain extra marks by correcting your own mistakes, using a better word and different grammatical features.

- Spelling and punctuation – all correct?
- Vocabulary – could you use a different word to make your writing more interesting? Have you used the same word too many times?
- Using 'and', 'but', 'then' – have you done so too many times? Maybe you need to use a full stop and capital letter instead.
- Sentences – have you used a variety of simple, compound and complex?
- Paragraphs – insert a square bracket ([) to indicate a new paragraph without rewriting.
- Tense – is it consistent throughout your personal recount?

Criteria:

There are ten criteria assessed in the writing task:

- audience
- text structure
- characters
- events
- vocabulary
- sentence structure
- paragraphs
- cohesion
- punctuation
- spelling.

9780170462198

TEST 1: Writing

MY HOLIDAY...

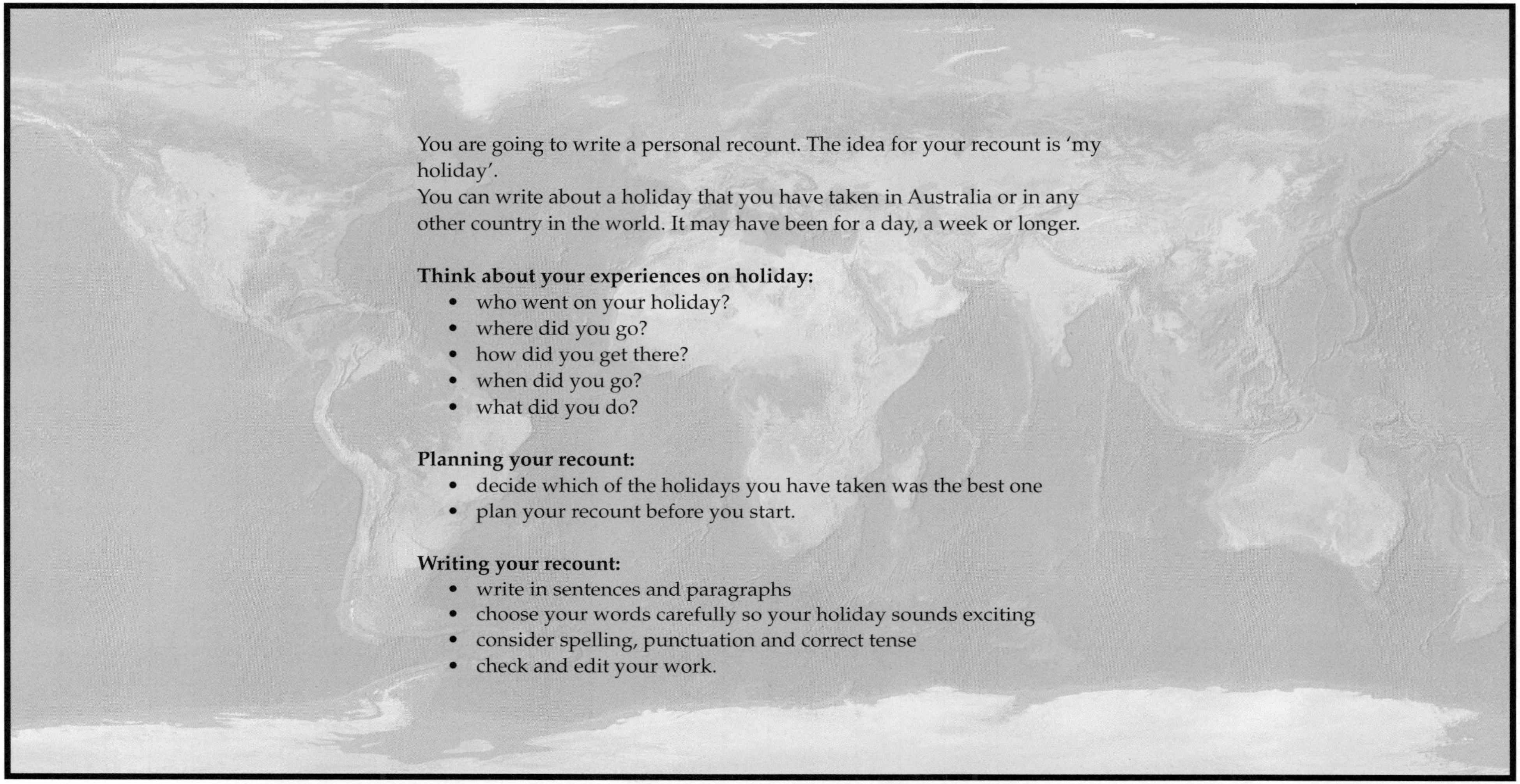

You are going to write a personal recount. The idea for your recount is 'my holiday'.
You can write about a holiday that you have taken in Australia or in any other country in the world. It may have been for a day, a week or longer.

Think about your experiences on holiday:

- who went on your holiday?
- where did you go?
- how did you get there?
- when did you go?
- what did you do?

Planning your recount:

- decide which of the holidays you have taken was the best one
- plan your recount before you start.

Writing your recount:

- write in sentences and paragraphs
- choose your words carefully so your holiday sounds exciting
- consider spelling, punctuation and correct tense
- check and edit your work.

Writing Test – Recount

Look at the stimulus. Brainstorm your ideas and then use this box to write a clear plan. Allow no more than five minutes. This planning will not be marked.

9780170462198

Year 7 Literacy

Language Conventions Test 2

Writing time: 45 minutes

Use 2B pencil only

Instructions

- Write your **student name** in the space provided.
- You must be silent during the test.
- If you need to speak to the teacher, raise your hand. Do not speak to other students.
- Answer all questions using a 2B pencil.
- If you wish to change your answer, erase it very thoroughly and then write your new answer.

Student name:

Each sentence has one word that is incorrect.
Write the correct spelling of the word in the box.

WRITE YOUR OWN ANSWER

1 The nawty dog had devoured my shoe.

2 The principle of the school addressed the assembly.

3 As quitely as a mouse, I crept up to the sleeping giant.

4 I always cry when I cut the layers of skin of an ungyun.

5 It is usefull to have a suitcase in which to keep your clothes when travelling.

6 I accidently forgot to brush my teeth before bed.

7 The busyness across the street closes before the fruit shop.

Each sentence has one word that is incorrect.
Write the correct spelling of the word in the box.

WRITE YOUR OWN ANSWER

8 I had to brake the glass to set off the fire alarm.

9 The accidents in the story are always caused by the main caracter.

10 The bewtiful princess was locked in a tower.

11 Glowball Warming has influenced the weather.

12 I have a freind who exercises regularly and plays basketball.

13 We enjoy laughing at humourous jokes.

14 It is essential to have your oxyjen tanks checked before scuba diving.

The spelling mistakes in these texts have been circled.
Write the correct spelling for each circled word in the box.

WRITE YOUR OWN ANSWER

15 The campaners in charge of the

riting of the information on the new

pamflet have been asked to create

posters to support their enviromental message.

9780170462198

16 It wasn't difficult to (reconise)

the dragon with its towering (presense,) the

(feirce) hot flames bursting from its terrifying mouth and

the (unusualy) large, sharp claws at the end of its gigantic paws.

17 (Febuary) is my favourite month,

because the (temprature) is so hot and

our cousins let us use (there) pool.

18 Which of the following correctly completes the sentence?

SHADE ONE BOX

I knew I _____ forget his name as it was also my brother's name.

☐ would've   ☐ would   ☐ wouldnt   ☐ wouldn't

19 Which of the following correctly completes the sentence?

SHADE ONE BOX

The little house was all aglow with its _____ Christmas lights wound round the verandah rails.

☐ kind   ☐ thoughtful   ☐ colourful   ☐ heavy

20 Which of the following correctly completes the sentence?

SHADE ONE BOX

That film about the alien encounter had the _____ creatures in it I had ever seen.

☐ weirdest   ☐ weirder   ☐ most weirdest   ☐ weird

21 Which of the following correctly completes the sentence?

SHADE ONE BOX

I was delighted that I had found my wallet after _____ everywhere for it.

☐ look   ☐ looks   ☐ had look   ☐ looking

22 Which of the following correctly completes the sentence?

SHADE ONE BOX

I [ ] the magic show before.

☐ had seen ☐ seen ☐ sees ☐ see

23 Which of the following correctly completes the sentence?

SHADE ONE BOX

C.S. Lewis, [ ] is the celebrated author of the *Narnia* series, would be amazed at the new film adaptations of his novels.

☐ who ☐ which ☐ what ☐ that

24 Which of the following correctly completes the sentence?

SHADE ONE BOX

I took my racquet hoping I would get to [ ] a game if the courts were available.

☐ played ☐ playing ☐ plays ☐ play

25 Which of the following correctly completes the sentence?

SHADE ONE BOX

I had worked [ ] than the others, because I paid closer attention to painting the exact detail of the creases and expression of the old woman's face.

☐ more slowly ☐ slowly ☐ slowest ☐ slow

26 Which sentence has the correct punctuation?

SHADE ONE BOX

☐ My favourite novels about Harry Potter have been made into films

☐ My favourite novels about Harry Potter have been made into films.

☐ My favourite novels about harry Potter have been made into films.

☐ My favourite novels about Harry potter have been made into films.

27 Which of the following correctly completes the sentence?

SHADE ONE BOX

Even though it was raining I chose not to catch the bus and [ ] all the way to school.

☐ walk ☐ walking ☐ walked ☐ walks

9780170462198

28 Which sentence is correct?

SHADE ONE BOX

- [ ] Myself and you are the only two who are prepared for this weather.
- [ ] You and myself are the only two who are prepared for this weather.
- [ ] Me and you are the only two who are prepared for this weather.
- [ ] You and I are the only two who are prepared for this weather.

29 Which of the following correctly completes the sentence?

SHADE ONE BOX

Do you have the autograph of any of ______ players in the winning team?

- [ ] them
- [ ] the
- [ ] its
- [ ] theirs

30 Which of the following correctly completes the sentence?

SHADE ONE BOX

While ______ patiently for the walk sign to turn green, the lady had time to put her mobile phone back into her bag.

- [ ] wait
- [ ] waiting
- [ ] waits
- [ ] waited

31 Which sentence has the correct punctuation?

SHADE ONE BOX

- [ ] Hurry and get on the bus before it rains said the teacher.
- [ ] 'Hurry and get on the bus before it rains,' said the teacher.
- [ ] 'Hurry and get on the bus,' before it rains said the teacher.
- [ ] 'Hurry and get on the bus before it rains said the teacher.'

32 I am ______ my jacket because rain has been predicted for this evening.

SHADE ONE BOX

- [ ] taking
- [ ] takes
- [ ] take
- [ ] taken

33 My brother's favourite thing to do after school is to ______ his bicycle through the park.

SHADE ONE BOX

- [ ] rode
- [ ] riding
- [ ] rides
- [ ] ride

34 Owing to the depth of the dive and the ______ of the water, the swimmer's goggles came off her face and she had to race without them.

SHADE ONE BOX

- [ ] impacting
- [ ] impact
- [ ] impacts
- [ ] impacted

35 Shade one box to show where the missing apostrophe (') should go.

SHADE ONE BOX

The farms☐ perimeter fences☐ and gates☐ were still in disrepair despite the farmers☐ trying hard to mend a section at a time.

36 Which sentence has the correct punctuation?

SHADE ONE BOX

- ☐ Mum said we were not to 'jump on the beds.'
- ☐ 'Mum said we were not to jump on the beds.'
- ☐ Mum said, 'We were not to jump on the beds.'
- ☐ Mum said we were not to jump on the beds.

37 Which sentence has the correct punctuation?

SHADE ONE BOX

- ☐ The seals and dolphins at the aquarium despite being fed by hand sometimes ate the fish living in their tank.
- ☐ The seals and dolphins, at the aquarium, despite being fed by hand sometimes ate the fish living in their tank.
- ☐ The seals and dolphins at the aquarium despite being fed by hand, sometimes ate the fish, living in their tank.
- ☐ The seals and dolphins at the aquarium, despite being fed by hand, sometimes ate the fish living in their tank.

38 Which sentence has the correct punctuation?

SHADE ONE BOX

- ☐ 'Watch your head said the tour guide.'
- ☐ Watch your head said the tour guide.
- ☐ 'Watch your head,' said the tour guide.
- ☐ Watch your head 'said the tour guide.'

39 Shade one box to show where the missing apostrophe (') should go.

SHADE ONE BOX

Red fire engines☐ flashed their lights and police cars☐ sirens☐ wailed to alert citizens☐ of their whereabouts.

40 Which sentence has the correct punctuation?

SHADE ONE BOX

- ☐ Dad said, 'it wasn't time to leave because the boat couldn't depart on a low tide.'
- ☐ Dad said it wasn't time to leave, 'because the boat couldn't depart on a low tide.'
- ☐ Dad said it wasn't time to leave, because the boat couldn't depart on a low tide.
- ☐ Dad said, 'it wasn't time to leave,' because the boat couldn't depart on a low tide.

41 Which sentence has the correct punctuation?

SHADE ONE BOX

- ☐ The man, wearing a striped shirt and tie who had recently started working for the department store was asked to serve, the waiting customers.
- ☐ The man wearing, a striped shirt and tie, who had recently started working for the department store was asked to serve the waiting customers.
- ☐ The man wearing a striped shirt and tie, who had recently started working for the department store, was asked to serve the waiting customers.
- ☐ The man wearing a striped, shirt and tie who had recently started working for the department store, was asked to serve the waiting customers.

42 Shade **two** boxes to show where the missing commas (,) should go.

SHADE TWO BOXES

Tickets ☐ for the final performance ☐ which sold out minutes ☐ after the phone lines were opened ☐ were sent ☐ out by mail.

Year 7 Literacy

Reading Magazine 2

TEXT 1

Bora Ring

Judith Wright

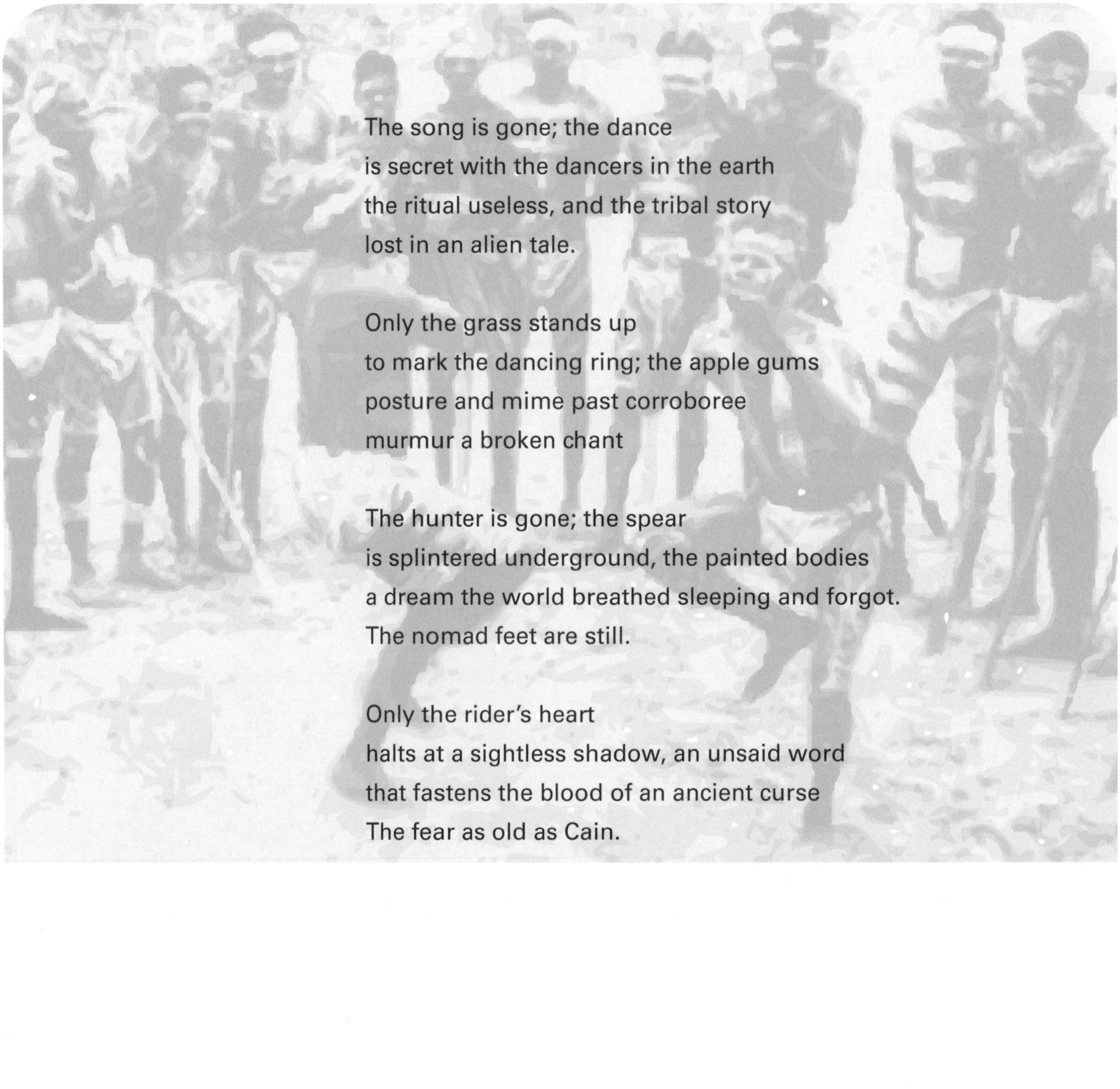

The song is gone; the dance
is secret with the dancers in the earth
the ritual useless, and the tribal story
lost in an alien tale.

Only the grass stands up
to mark the dancing ring; the apple gums
posture and mime past corroboree
murmur a broken chant

The hunter is gone; the spear
is splintered underground, the painted bodies
a dream the world breathed sleeping and forgot.
The nomad feet are still.

Only the rider's heart
halts at a sightless shadow, an unsaid word
that fastens the blood of an ancient curse
The fear as old as Cain.

9780170462198

Ancient Egypt

From Wikipedia, the online free encyclopedia where contributors are free to write what they want.

The pyramids are among the most recognizable symbols of the civilization of ancient Egypt.

Ancient Egypt was an ancient civilization in eastern North Africa, concentrated along the lower reaches of the Nile River in what is now the modern nation of Egypt. The civilization began around 3150 BC with the political unification of Upper and Lower Egypt under the first pharaoh, and it developed over the next three millennia. Its history occurred in a series of stable periods, known as kingdoms, separated by periods of relative instability known as Intermediate Periods. After the end of the last kingdom, known as the New Kingdom, the civilization of ancient Egypt entered a period of slow, steady decline, during which Egypt was conquered by a succession of foreign powers. The rule of the pharaohs officially ended in 31 BC when the early Roman Empire conquered Egypt and made it a province.

The civilization of ancient Egypt thrived from its ability to adapt to the conditions of the Nile River Valley. Controlled irrigation of the fertile valley produced surplus crops, which fueled social development and culture. With resources to spare, the administration sponsored mineral exploitation of the valley and surrounding desert regions, the early development of an independent writing system, the organization of collective construction and agricultural projects, trade with surrounding regions, and a military that defeated foreign enemies and asserted Egyptian dominance. Motivating and organizing these activities was a bureaucracy of elite scribes, religious leaders, and administrators under the control of a pharaoh who ensured the cooperation and unity of the Egyptian people through an elaborate system of religious beliefs.

The many achievements of the ancient Egyptians included a system of mathematics, quarrying, surveying and construction techniques that facilitated the building of monumental pyramids, temples, obelisks, faience and glass technology, a practical and effective system of medicine, new forms of literature, irrigation systems and agricultural production techniques, and the earliest known peace treaty. Egypt left a lasting legacy: art and architecture were copied and antiquities paraded around the world.

Cyber Bullying

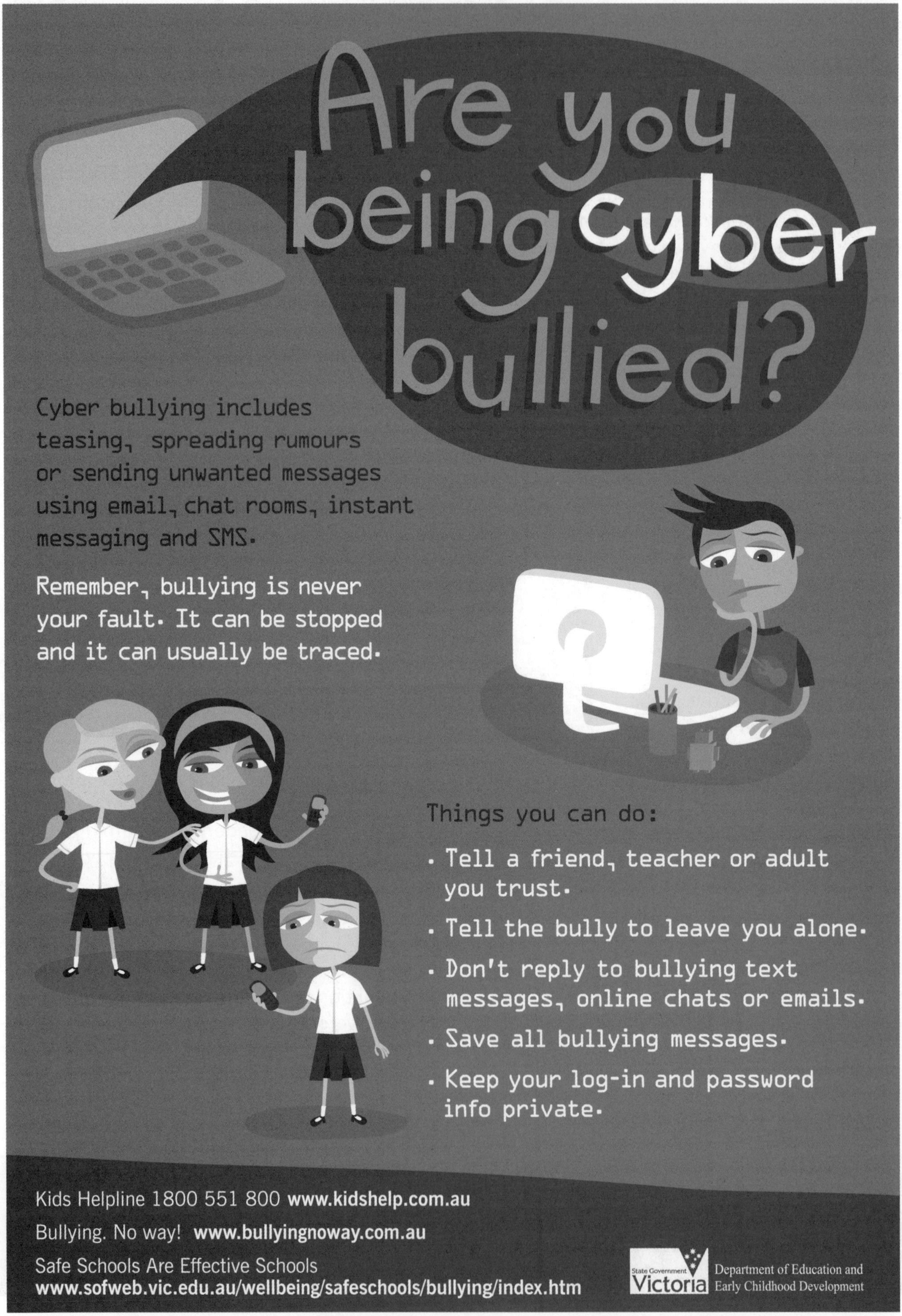
Are you being cyber bullied?
Cyber bullying includes teasing, spreading rumours or sending unwanted messages using email, chat rooms, instant messaging and SMS.
Remember, bullying is never your fault. It can be stopped and it can usually be traced.
Things you can do:
• Tell a friend, teacher or adult you trust.
• Tell the bully to leave you alone.
• Don't reply to bullying text messages, online chats or emails.
• Save all bullying messages.
• Keep your log-in and password info private.
Kids Helpline 1800 551 800 www.kidshelp.com.au
Bullying. No way! www.bullyingnoway.com.au
Safe Schools Are Effective Schools
www.sofweb.vic.edu.au/wellbeing/safeschools/bullying/index.htm
State Government Victoria
Department of Education and Early Childhood Development

9780170181754

Understanding Maps

Maps can be used to provide information about many different things. We can use mind maps or graphic organisers to summarise information. There are weather maps that can inform us about today's weather or predict weather forecasts days and weeks in advance.

Maps can also provide us with directions such as roads, tourist attractions, and campgrounds. These can be in print form such as a street directory or electronic form such as an electronic navigation tool or a mobile phone.

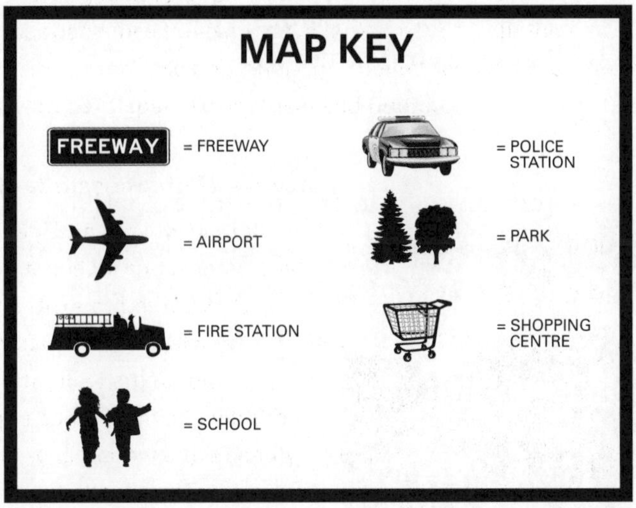

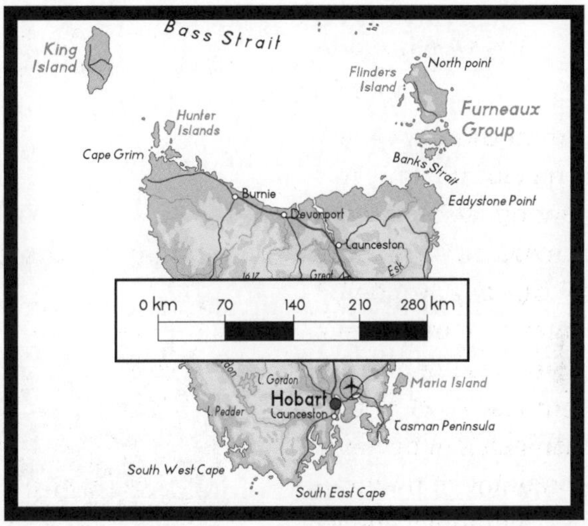

Symbols are used on a map to represent objects and places of interest. A symbol is a picture that is used to represent something in the real world. Understanding these symbols requires the use of a key. These keys usually show a small picture of each of the symbols used on the map, along with a written description of the meaning for each of these symbols. This is known as a key or legend for each of the symbols used in the map.

A map's scale is the connection between the map distance and the distance in the real world. Maps are created to scale. This is so precise distances can be calculated from something much smaller than the actual area covered by a map. A Graphic Scale is one method used to measure the distance between two objects on a map, using a line, with divisions marked by smaller intersecting lines, similar to markers on a ruler. One side of the scale represents the distance on the map; the other side represents the exact distances of objects in real life. By measuring the distance between two objects on a map, and then referring to the graphic scale, it is easy to calculate the actual distance between those same points on a map.

There are three common methods used by map makers to depict scale. These methods are referred to as the graphic method, the verbal method, and the fractional method.

Map keys and legends can be found in ancient records as well as modern documents. Map scales have become more accurate with modern technology measuring distances from space tracking stations. Some e-maps are able to show minute details about an area as well as geographical changes as they happen.

The Silver Brumby

Elyne Mitchell

Down near the Cascades hut, there were some tall trees, candlebarks and the first of a great mountain ash, and the two foals had already discovered the fun of playing 'tug-you-last' around the great tree-trunks and up and down clear the glades. Now, as soon as they were in the timber they could hear the wail of the wind in the tree-tops, far above, and the soughing and sighing of streamers of bark that hung down the trunks.

They felt very small and alone – and very excited.

'What was that?' asked Thowra nervously, as something white and feathery floated down from the dark sky and landed, freezing cold, on his nose.

Storm jumped to one side and shook his head as another cold white feather fell on his ear. They cantered away under a big tree, but, even there, floating so slowly and lightly on the air, the white feathers came, in ones and twos at first, but thicker and thicker until the air was filled with floating whiteness.

It was a long time before they thought of looking at the ground.

'Look!' cried Storm. 'It's even making the ground white. We should go home – it might be difficult to find our way if we don't leave now.'

They had no trouble finding their way back while they were in among the trees and the trunks to guide them, but the open valley was a blinding whirl of blown whiteness. The shape of the tracks could still be seen, and Thowra jogged along one, his nose to the ground. Storm ran right beside him, almost bumping him.

'You'll tread on me,' Thowra complained. 'What is the matter?'

'I can hardly see you through this strange white stuff,' said Storm, and he sounded afraid. His own dark coat showed up clearly, but Thowra was almost invisible.

The Duke of Edinburgh's Award

Tasmania

The Duke of Ed

How it all began...

Welcome to The Duke of Edinburgh's Award in Tasmania. The Award has been operating in Tasmania since October 1962 and has a strong history of participation. Although the basic philosophy of The Award has not changed since the early days it has moved with the times, its more flexible, more accessible, more exciting. There has never been a better time to get involved, either as an Award participant or as a licensed operator.

A great place to do The Award...

Tasmania has many natural advantages for those interested in The Duke of Edinburgh's Award. Our bush walking opportunities are unique. People come from all over the world to experience the beauty and the challenge of Tasmania's outdoors. Add to this access to lakes, rivers and the sea with all the accompanying networks and we have a very exciting choice for our Adventurous Journeys.

Our communities have built effective systems that offer access to any number of sports on a regular basis often with high level coaching and assistance. There are many sports readily available to those keen enough to seek them out. Every country community has it's own teams and opportunities that are part of the wider sporting community. How far you want to go is up to you. Just check out, through your local council website, lists of activities that are available now.

These same communities provide the network for Volunteering which gives you the chance to give something back from which you will gain experience and satisfaction. Tasmania relies on volunteer input for almost everything and young people with a sense of community are in demand.

Alongside Physical Activities there are opportunities in Tasmania to acquire Skills from a wide variety of sources. For example, music, art or technical skills like electronics or horse riding. All it takes is the desire to be involved. You are only limited by your imagination.

We are very fortunate to live in Tasmania, let's make the most of it.

About the Award: History

The Award program was first introduced in the United Kingdom in 1956 as The Duke of Edinburgh's Award. The aim was to motivate boys aged between 15 and 18 to become involved in a balanced program of voluntary self-development activities to take them through the potentially difficult period between adolescence and adulthood.

A girls' scheme was launched in 1958, and the two separate schemes were amalgamated in 1969. In 1957 the upper age limit was increased to 19, increased again in 1965 to 20, increased to 21 in 1969, and finally increased to 25 in 1980.

The Girl with No Name

Pat Lowe

On a broader ledge about halfway up the valley wall stood a young boab tree. It seemed to be doing very well in spite of its perilous and infertile position. Trees clung to other ledges, too small to hold them. Their exposed roots went down to the rocks to find a soil-filled niche somewhere.

Matthew unzipped his backpack and pulled out a sandwich, devouring it hungrily. An ant, big as a beetle, discovered the crumbs he let fall, and raced off with one of them held triumphantly between its mandibles.

Matthew knelt by the pool and scooped a few handfuls of water to his mouth. As he straightened up, his mind suddenly relaxed, free of thought, allowing his senses to be filled by the nature around him until, momentarily, he became part of it. Then his self-awareness returned, and with it his feeling of purpose. It was time to start looking for rock paintings.

First, he stowed his pack and sleeping bag in the crevice between two rocks, where they were hidden from sight by long grass and wattle bushes. He clipped his water bottle onto his belt, stuffed an apple in one pocket and a packet of dried fruit and nuts in the other, and started off.

The valley was dry, but Matthew headed downstream from the pool, where water flowed during the rainy season. As he picked his way over the boulders he kept his eye on the western wall of the valley, looking for likely caves or galleries where the paintings might be. From time to time he glanced over towards the eastern wall, hoping to recognise the boab tree that would give him his bearings.

9780170462198

TEXT 8

Puffing Billy Australia's Favourite Steam Train

A guide to your journey

Welcome to Puffing Billy!

The Puffing Billy Railway welcomes you aboard the train for a trip through the scenic Dandenong Ranges. We trust that your journey will be an enjoyable and memorable experience.

What is Puffing Billy?

The Puffing Billy railway was one of four low-cost 762mm (2'6") gauge lines constructed in Victoria in the early 1900s to open up remote areas. The present line between Belgrave and Gembrook, through the forests, fern gullies and farmlands of the magnificent Dandenong Ranges, is the major part of the line which opened on 18 December 1900 and operated over 29km (18.2 miles) between Upper Ferntree Gully and Gembrook until 1953. In 1953, a landslide blocked the track and, because of operating losses, the line was closed.

Public interest resulted in the formation of the Puffing Billy Preservation Society, whose volunteers, with the blessings of the Victorian Railways and the assistance of the Citizens' Military Forces, by-passed the landslide and reopened the line to Menzies Creek in 1962, Emerald in 1965, Lakeside in 1975 and finally to Gembrook in October 1998.

Puffing Billy Today

Now the goods and livestock have gone but the passengers have returned in greater numbers than ever. Today Puffing Billy is Australia's favourite steam train and one of the finest preserved steam railways in the world. It operates every day except Christmas Day, thanks to the tireless efforts of more than 600 dedicated volunteers.

Group bookings on all scheduled trains may be made and special charter trains for groups of up to 250 are available. 'Special Interest' groups can also be catered for with escorted tours.

Puffing Billy trains depart from Belgrave, only 40 kms or one hour east of Melbourne by car or coach. Easy access is also available by electric train from Melbourne in around seventy minutes, with a short walk from the electric train station to the Puffing Billy station.

Day Tours

The railway offers a range of stand-alone experiences or is available through one of the excellent day tours operated seven days a week by AAT Kings, Australian Pacific Tours, Gray Line Melbourne or Great Sights.

These day tours package the railway with a number of the other major tourist attractions on the eastern side of Melbourne, including the world famous Penguin Parade at Phillip Island, Healesville Sanctuary or a selection from the many fine wineries of the Yarra Valley.

Dining Trains

Puffing Billy offers a range of unique first class 'Wine and Dine' experiences in luxury, fully enclosed heritage dining carriages. Travel on the daily 'Steam & Cuisine' luncheon train or 'The Devonshire Journey' afternoon tea train. Alternatively, on Fridays and Saturdays you can enjoy the evening 'Dinner Special'. Charter a dining train for that special event, whether it is a wedding, corporate function or social occasion.

Bookings are essential.

'Day Out with Thomas'

Puffing Billy is pleased to welcome 'Thomas the Tank Engine' to our Emerald Town Station during Autumn and Spring. Children and adults can see a real 'Thomas' in steam, meet the Fat Controller, enjoy a special pantomime and take a steam train ride. Bookings are essential.

General Information

DISABLED PASSENGERS Specially designed carriages accommodate a limited number of passengers in wheelchairs. Please phone to check availability.

SMOKING Smoking is not permitted on Puffing Billy trains or in railway buildings.

REFRESHMENTS Refreshment Rooms at main stations stock snacks, confectionery, drinks, ice creams and souvenirs, including souvenir publications about the history of Puffing Billy.

LOST PROPERTY Lost property should be reported to our Station Master at Belgrave. Any property found on the Railway should be handed to the Station Master at Belgrave or to any Railway official at other stations.

TIMETABLES For 24-hour recorded timetable and fare information, phone (61 3) 9757 0700.

Enquiries & Bookings

For general enquiries and bookings, please contact our Belgrave office during business hours, Monday to Friday:

Phone: (61 3) 9757 0700 Fax: (61 3) 9757 0705

Email: INFO@PBR.ORG.AU

Puffing Billy Railway

P.O. Box 451, Belgrave 3160, Australia.

WWW.PUFFINGBILLY.COM.AU

Published by Emerald Tourist Railway Board.

06/09

Australia's Favourite Steam Train

Runs every day except Christmas Day!

WWW.PUFFINGBILLY.COM.AU

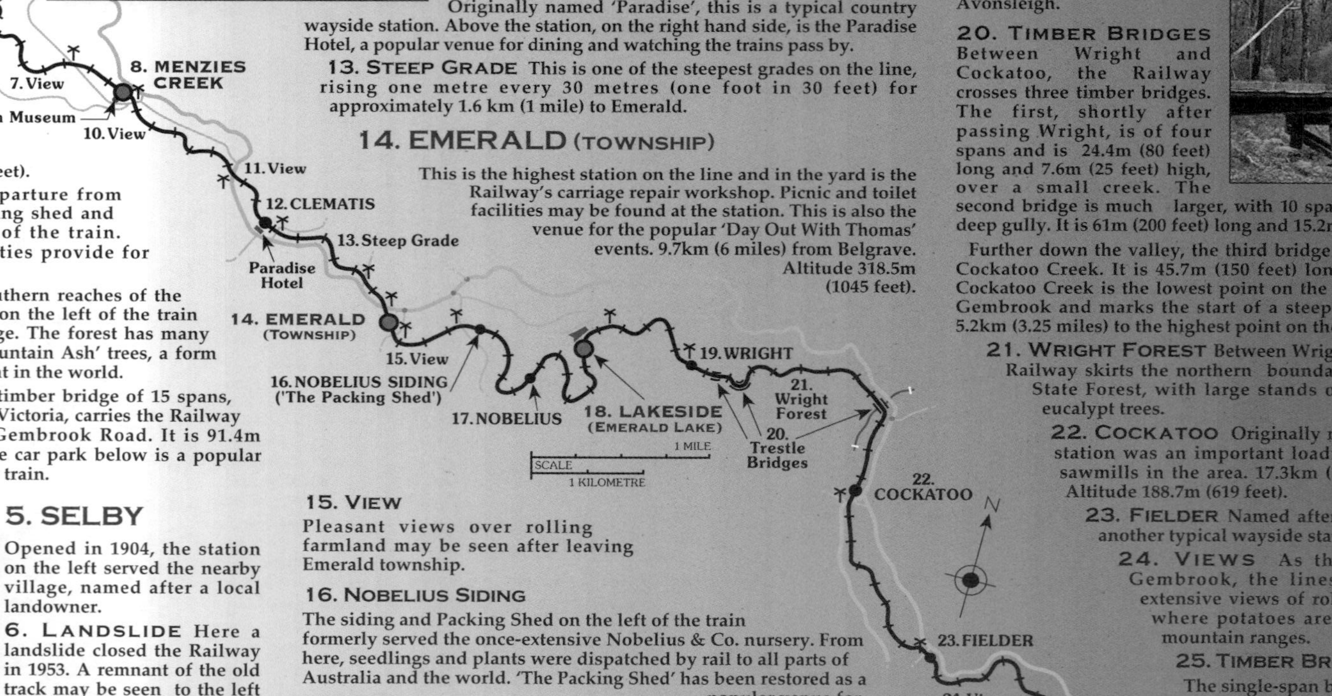

1. BELGRAVE

This station is the headquarters of the Railway, with both operating and administrative facilities. Most Puffing Billy trains commence their journeys here. Rail distance from Melbourne 41.8km (26 miles). Altitude 227.7m (747 feet).

2. LOCOMOTIVE DEPOT On departure from Belgrave station, the locomotive running shed and workshops may be seen to the left of the train. Extensive servicing and repair facilities provide for continuous maintenance and restoration.

3. SHERBROOKE FOREST The southern reaches of the forest extend to the Puffing Billy track on the left of the train and to the area beside the Trestle Bridge. The forest has many magnificent examples of the native 'Mountain Ash' trees, a form of eucalypt and the tallest flowering plant in the world.

4. TRESTLE BRIDGE This famous timber bridge of 15 spans, now classified by the National Trust of Victoria, carries the Railway over Monbulk Creek and the Main Gembrook Road. It is 91.4m (300 feet) long, 12.8m (42 feet) high. The car park below is a popular place for viewing and photographing the train.

5. SELBY

Opened in 1904, the station on the left served the nearby village, named after a local landowner.

6. LANDSLIDE Here a landslide closed the Railway in 1953. A remnant of the old track may be seen to the left of the train. Before the line was closed, a water tank for the locomotives was located here.

7. VIEW An extensive view to Port Phillip Bay is to be seen to the right of the train on the approach to Menzies Creek.

8. MENZIES CREEK

Named after an early miner who worked in the area. Trains in opposite directions often 'cross' here. 6km (3.75 miles) from Belgrave. Altitude 303.6 m (996 feet).

9. STEAM MUSEUM Adjacent to Menzies Creek station, the Steam Museum houses a unique collection of locomotives, rolling stock, steam machinery and other rare items from the 'steam era' of bygone years. The Steam Museum is currently closed for redevelopment and will re-open at a later date after works are completed.

10. VIEW On a clear day, extensive views to Port Phillip Bay, Arthurs Seat and Westernport may be seen to the right of the train. Also on the right and below is the Cardinia Reservoir, which provides water to the south-eastern suburbs of Melbourne.

11. VIEW After crossing the Main Road, the train enters Paradise Valley, with pleasant farmland views. Kiwi fruit are grown on the vine plantation to the left of the train near Clematis station.

12. CLEMATIS

Originally named 'Paradise', this is a typical country wayside station. Above the station, on the right hand side, is the Paradise Hotel, a popular venue for dining and watching the trains pass by.

13. STEEP GRADE This is one of the steepest grades on the line, rising one metre every 30 metres (one foot in 30 feet) for approximately 1.6 km (1 mile) to Emerald.

14. EMERALD (TOWNSHIP)

This is the highest station on the line and in the yard is the Railway's carriage repair workshop. Picnic and toilet facilities may be found at the station. This is also the venue for the popular 'Day Out With Thomas' events. 9.7km (6 miles) from Belgrave. Altitude 318.5m (1045 feet).

15. VIEW

Pleasant views over rolling farmland may be seen after leaving Emerald township.

16. NOBELIUS SIDING

The siding and Packing Shed on the left of the train formerly served the once-extensive Nobelius & Co. nursery. From here, seedlings and plants were dispatched by rail to all parts of Australia and the world. 'The Packing Shed' has been restored as a popular venue for 'wine & dine' functions such as weddings and the evening 'Dinner Specials'.

10.2km (6.4 miles) from Belgrave.

17. NOBELIUS

Named after the former Nobelius estates, this small wayside station on the right of the train is almost half way between Belgrave and Gembrook.

18. LAKESIDE (EMERALD LAKE)

This was the terminus of the line before reopening to Gembrook in 1998. Trains usually pause here for the crew to fill the locomotive water tanks. Picnic and barbecue facilities are in the adjacent Emerald Lake Park, which offers pleasant walks, a pool and paddle boats in the summer months. Toilets are provided at the station. 13.2km (8.25 miles) from Belgrave. Altitude 242m (795 feet).

19. WRIGHT On the left of the train, this was formerly the station for the town of Avonsleigh.

20. TIMBER BRIDGES Between Wright and Cockatoo, the Railway crosses three timber bridges. The first, shortly after passing Wright, is of four spans and is 24.4m (80 feet) long and 7.6m (25 feet) high, over a small creek. The second bridge is much larger, with 10 spans, carrying the line over a deep gully. It is 61m (200 feet) long and 15.2m (50 feet) high.

Further down the valley, the third bridge has 10 straight spans, over Cockatoo Creek. It is 45.7m (150 feet) long and 4.6m (15 feet) high. Cockatoo Creek is the lowest point on the line between Belgrave and Gembrook and marks the start of a steep (1 in 30) uphill climb for 5.2km (3.25 miles) to the highest point on the line near Gembrook.

21. WRIGHT FOREST Between Wright and Cockatoo Creek, the Railway skirts the northern boundary of the un-spoilt Wright State Forest, with large stands of Stringy Bark and other eucalypt trees.

22. COCKATOO Originally named Cockatoo Creek, the station was an important loading point for timber from sawmills in the area. 17.3km (10.8 miles) from Belgrave. Altitude 188.7m (619 feet).

23. FIELDER Named after local land-owners, this is another typical wayside station on the left of the train.

24. VIEWS As the train climbs towards Gembrook, the lineside bush gives way to extensive views of rolling cultivated farmlands, where potatoes are still grown, and distant mountain ranges.

25. TIMBER BRIDGE

The single-span bridge carries the Railway over an old farm access track, now disused. Shortly after the bridge, the train climbs one of the longest straight sections of the line.

26. GEMBROOK

Once a busy centre for the dispatch of timber and farm produce from the district. Picnic facilities are provided at the station and in the adjacent park. Trains usually stop over for an hour or more for visitors to explore the historic town before the return trip. 24km (15 miles) from Belgrave. Altitude 311.8m (1020 ft).

9780170462198

Year 7 Literacy

Reading Test 2

Writing time: 65 minutes

Use 2B pencil only

Instructions

- Write your **student name** in the space provided.
- You must be silent during the test.
- If you need to speak to the teacher, raise your hand. Do not speak to other students.
- Answer all questions using a 2B pencil.
- If you wish to change your answer, erase it very thoroughly and then write your new answer.

Student name:

Text 1: *Bora Ring* Questions

Shade one box to show the correct answer to the following questions.

1 Judith Wright is an Australian Poet. 'Bora Ring' is a poem about

SHADE ONE BOX

☐ Aboriginal traditions.
☐ a volcano erupting.
☐ dancing outside.
☐ Aboriginal people.

2 What is a Bora Ring?

SHADE ONE BOX

☐ An Australian Aborigine's wedding ring.
☐ A symbol on a map key.
☐ A cultural site important to Indigenous Australians.
☐ The boundary of an important farming area.

3 A corroboree is

SHADE ONE BOX

☐ a traditional dancing, ceremonial meeting of Australian Aborigines.
☐ a home made from bark and leaves.
☐ a musical instrument played by Indigenous Australians.
☐ a display at the Australian Museum.

4 The repetition of line length helps to develop

SHADE ONE BOX

☐ discord.
☐ rhyme.
☐ pattern.
☐ rhythm.

5 How does the writer want us to feel?

SHADE ONE BOX

☐ Sad because the Bora Ring is gone.
☐ Relieved because it was too noisy.
☐ Angry because it was vandalised.
☐ Unhappy and wants us to empathise.

9780170462198

Text 2: *Ancient Egypt* Questions

Shade one box to show the correct answer to the following questions.

1 *Wikipedia* may not be a reliable source of information because

SHADE ONE BOX

☐ it is a primary source.
☐ you have to pay to read it.
☐ encyclopaedias are only correct as a book.
☐ it is an online encyclopaedia that anyone can edit.

2 According to the text where was Ancient Egypt?

SHADE ONE BOX

☐ close to Lower Egypt
☐ the Nile River Delta
☐ North Africa
☐ Eastern North Africa

3 The pharaohs ruled ancient Egypt for approximately __________ years.

SHADE ONE BOX

☐ 30 ☐ 100 ☐ 1000 ☐ 3000

4 Egyptian scribes developed

SHADE ONE BOX

☐ handwriting classes for students.
☐ an independent writing system.
☐ graffiti art on pyramid walls.
☐ the A – Z alphabet.

5 According to the text Ancient Egypt's success was mainly due to

SHADE ONE BOX

☐ the support of the people.
☐ the strength of their armies.
☐ successful farming and trade.
☐ the ability of construction teams.

6 The rule of the Pharaohs came to an end when

SHADE ONE BOX

☐ Upper and Lower Egypt became joined.
☐ they were overpowered by the Romans.
☐ the Nile River changed its course.
☐ their military defeated other armies.

Text 3: *Cyber Bullying* Questions

Shade one box to show the correct answer to the following questions.

1 According to the text what does the word cyber mean?

SHADE ONE BOX

- ☐ It describes a person who misuses the Internet.
- ☐ It is the name for a computer hacker.
- ☐ It describes a person, living in the computer and information age.
- ☐ It is a person who takes care of computers.

2 According to the text, what is one way you can protect yourself against cyber bullies?

SHADE ONE BOX

- ☐ Nothing, you just have to put up with it.
- ☐ Never give anyone your password and log-in details.
- ☐ Get your friends to bully them back.
- ☐ Turn off your computer.

3 SMS messages that are trying to bully anyone should be

SHADE ONE BOX

- ☐ deleted immediately.
- ☐ forwarded to your friends.
- ☐ replied to so the person knows you got it.
- ☐ saved and shown to a trusted adult.

4 According to the text what should you remember?

SHADE ONE BOX

- ☐ Being bullied is not your fault.
- ☐ Always turn off your computer.
- ☐ You can never stop bullying.
- ☐ Always fight bullying with bullying.

5 Why does the poster tell you to 'Save all bullying messages'?

SHADE ONE BOX

- ☐ so the writer can be identified.
- ☐ so you can write back to the bully.
- ☐ so people know you are telling the truth.
- ☐ so you can forward copies to other people.

9780170462198

Text 4: *Understanding Maps*
Questions

Shade one box to show the correct answer to the following questions.

SHADE ONE BOX

1 Mind maps can be used to

☐ make notes.

☐ measure distance.

☐ predict weather.

☐ gauge intelligence.

SHADE ONE BOX

2 According to the text, a 'place of interest' might be a

☐ church.

☐ school.

☐ ambulance.

☐ hospital.

SHADE ONE BOX

3 A symbol is used in a map key to show

☐ real places.

☐ how far between towns.

☐ today's weather.

☐ traffic lights

SHADE ONE BOX

4 Choose the correct words to complete this sentence. Map scales are used to extract ____________________ ____________________ shown on a map.

☐ approximate distances

☐ exact distances

☐ approximate direction

☐ exact direction

SHADE ONE BOX

5 According to the text 'the fractional method' can be used to

☐ place grid lines on a map.

☐ represent a scale on a map.

☐ insert compass points on a map.

☐ identify axis points on a map.

SHADE ONE BOX

6 Modern maps are more accurate than ancient ones because

☐ we can use satellites to develop more accurate maps.

☐ cars have reliable speedos to measure distance accurately.

☐ there are a lot more people making maps now.

☐ maps can be made by tracking hand-held devices.

Text 5: *The Silver Brumby* Questions

Shade one box to show the correct answer to the following questions.

1 What is another word for cascade?

SHADE ONE BOX

- ☐ tree
- ☐ waterfall
- ☐ river
- ☐ stoney

2 Thowra and Storm are

SHADE ONE BOX

- ☐ dogs.
- ☐ birds.
- ☐ horses.
- ☐ people.

3 The effect of using 'the wail of the wind in the treetops' is to

SHADE ONE BOX

- ☐ explain how tall the trees are.
- ☐ emphasise how scared the horses were.
- ☐ describe the different types of trees.
- ☐ personify the sounds of nature.

4 The floating white feathers were

SHADE ONE BOX

- ☐ sleet.
- ☐ snowflakes.
- ☐ flower petals.
- ☐ cockatoo feathers.

5 According to the text Storm sounded afraid. List one of his actions that shows he is afraid.

WRITE YOUR OWN ANSWER

____

6 The colour of Thowra's coat is most likely

SHADE ONE BOX

- ☐ white.
- ☐ black.
- ☐ brown.
- ☐ patchy.

9780170462198

Text 6: *Duke of Edinburgh Awards in Tasmania* Questions

Shade one box to show the correct answer to the following questions.

1 The Duke of Edinburgh Award was introduced for

SHADE ONE BOX

- ☐ boys who lived in Tasmania.
- ☐ more Aussie kids to have a go at sport.
- ☐ anyone who was under the age of 18 years.
- ☐ boys over the age of 14 and under 18 years.

2 According to the text the 'the two separate schemes were amalgamated in 1969' meant that

SHADE ONE BOX

- ☐ girls were allowed to participate.
- ☐ boys and girls could earn the same awards.
- ☐ Tasmania joined the rest of Australia to host the awards.
- ☐ 'there were many more participants'.

3 According to the text the 'the basic philosophy of The Award' refers to the

SHADE ONE BOX

- ☐ successful personal growth of young people through volunteering.
- ☐ exciting adventures that families can do together in Tasmania.
- ☐ need for more young people to have healthier lifestyles.
- ☐ encouraging of parents to volunteer at community events.

4 According to the text why do people from all over the world choose to visit Tasmania?

SHADE ONE BOX

- ☐ It is a cheap place to go for a holiday.
- ☐ To enjoy the experience of bush walking in a picturesque place.
- ☐ It is the only place in the world you can do the Duke of Edinburgh Award.
- ☐ It is only a small island so it's well contained.

5 According to the text to find out about local community sports you should

SHADE ONE BOX

- ☐ read the newspapers everyday.
- ☐ join a lot of different clubs.
- ☐ look up a local council website.
- ☐ search the Duke of Edinburgh website.

6 This photograph has most likely been chosen to show us that

SHADE ONE BOX

- ☐ cars can only travel in certain parts of Tasmania.
- ☐ some places in Tasmania are very striking.
- ☐ there are a variety of activities to join in on Tasmania.
- ☐ living in Tasmania might be lonely.

7 Young people with a sense of community are valued because

SHADE ONE BOX

- ☐ Tasmania relies on volunteers.
- ☐ Tasmania has a lot jobs available.
- ☐ Tasmania needs more people to coach awardees.

Text 7: *The Girl with No Name* Questions

Shade one box to show the correct answer to the following questions.

1 According to the text the boab tree was

SHADE ONE BOX

- ☐ near the pool.
- ☐ down in the valley.
- ☐ where Matthew hid his backpack.
- ☐ growing in a dangerous place.

2 The writer uses a metaphor 'big as a beetle' to describe the size of the ant as

SHADE ONE BOX

- ☐ smaller than most ants.
- ☐ taller than most ants.
- ☐ being about the same size.
- ☐ much larger than most ants.

3 When the ant took a crumb Matthew dropped it was

SHADE ONE BOX

- ☐ scared.
- ☐ proud.
- ☐ unhappy.
- ☐ happy.

4 According to the text when Matthew 'stowed his pack and sleeping bag in the crevice' it is telling us he

SHADE ONE BOX

- ☐ put it under the boab tree in the shade.
- ☐ carefully put it in a hole in the rocks.
- ☐ threw it on the ground and left it there.
- ☐ accidentally dropped it in the pool of water.

5 According to the text the stream had no

SHADE ONE BOX

- ☐ trees.
- ☐ water.
- ☐ paintings.
- ☐ sunlight.

9780170462198

6 Why is the boab tree important to Matthew?

WRITE YOUR OWN ANSWER

Explain your answer.

____

____

____

____

____

7 How does Matthew feel?

SHADE ONE BOX

- ☐ cautious
- ☐ intrigued
- ☐ determined
- ☐ overwhelmed

Text 8: *Puffing Billy* Questions

Shade one box to show the correct answer to the following questions.

1 The most important station for the Puffing Billy train is at

SHADE ONE BOX

- ☐ Gembrook.
- ☐ Nobelius Siding.
- ☐ Belgrave.
- ☐ Menzies Creek.

2 Select the correct words to complete this sentence. You can see all the way to ____________________ on a sunny day.

SHADE ONE BOX

- ☐ Arthurs Seat
- ☐ a large landslide
- ☐ Emerald Lake Park
- ☐ Thomas the Tank Engine

3 Which station has the highest altitude on the Puffing Billy train ride?

SHADE ONE BOX

- ☐ Fielder
- ☐ Selby
- ☐ Wright
- ☐ Emerald

4 According to the text at 'The Packing Shed' people can

SHADE ONE BOX

- ☐ change trains.
- ☐ get married.

☐ have a picnic.

☐ go for a swim.

5 The trestle bridges are either side of which area?

SHADE ONE BOX

☐ Wright Forest

☐ Port Phillip Bay

☐ Cardinia Reservoir

☐ Sherbrooke Forest

6 What is the effect of including pictures and a map in the layout of this text?

WRITE YOUR OWN ANSWER

7 The bridges along the railway line are made from

SHADE ONE BOX

☐ steel.

☐ timber.

☐ trestles.

☐ planks.

9780170462198

Year 7 Literacy

Writing Test 2

Writing time: 40 minutes

Use 2B pencil, blue or black pen only

Instructions

- Write your **student name** in the space provided.
- You must be silent during the test.
- If you need to speak to the teacher, raise your hand. Do not speak to other students.
- Use a 2B pencil or a black or blue pen only.
- Use the lines provided. Do not write in the borders.

Student name:

You are going to write a narrative.

The idea for your narrative is 'friendship'.

- Read through the ideas listed on the stimulus.
- Choose the ideas that appeal to you.
- You don't have to use all of the ideas provided. They are offered to get you thinking quickly and assist in composing in a short, specific time frame.

Your own idea:

____

____

____

____

____

____

Planning your narrative:

- think about your characters. (Give your characters names, personalities and maybe a description.)
- where are they? (This is the setting.)
- make sure that the main characters and/or setting are included in the introduction.
- where are they going and how will they get there? (This is part of the sequence of events that forms the body of your story.)
- dialogue markers maybe necessary if the characters speak to one another
- think about your ending. How can you make it exciting, surprising or entertaining?

Crafting your narrative:

- use interesting and emotive words (include multi-syllable words and try not to be repetitive in your choice of words).
- use great action verbs (for example, think of other words to use instead of 'said' – such as 'whispered', 'stammered', 'screamed').
- spelling (be careful to spell words correctly).
- punctuation (capital letters, full stops, commas, speech markers – all the required punctuation to assist the flow and cohesion of your narrative).
- correct tense (maintain the same tense throughout your story; decide whether it is past, present or future).

Writing your story:

- use a clear structure to build a cohesive and engaging narrative.
- use sentences that include a variety of simple, compound and complex and use correct punctuation.
- use paragraphs that begin with a topic sentence that introduces and develops a key thesis or central idea.

Checking your work:

Always read back over your writing to check for opportunities to gain extra marks by correcting your own mistakes, using a better word and different grammatical features.

- Spelling and punctuation – all correct?
- Vocabulary – could you use a different word to make your writing more interesting? Have you used the same word too many times?
- Using 'and', 'but', 'then' – have you done so too many times? Maybe you need to use a full stop and capital letter instead.
- Sentences – have you used a variety of simple, compound and complex?
- Paragraphs – insert a square bracket ([) to indicate a new paragraph without rewriting.
- Tense – is it consistent throughout your narrative?

Criteria:

There are ten criteria assessed in the writing task:

- audience
- text structure
- characters
- events
- vocabulary
- sentence structure
- paragraphs
- cohesion
- punctuation
- spelling.

9780170462198

TEST 2: Writing

FRIENDSHIP…

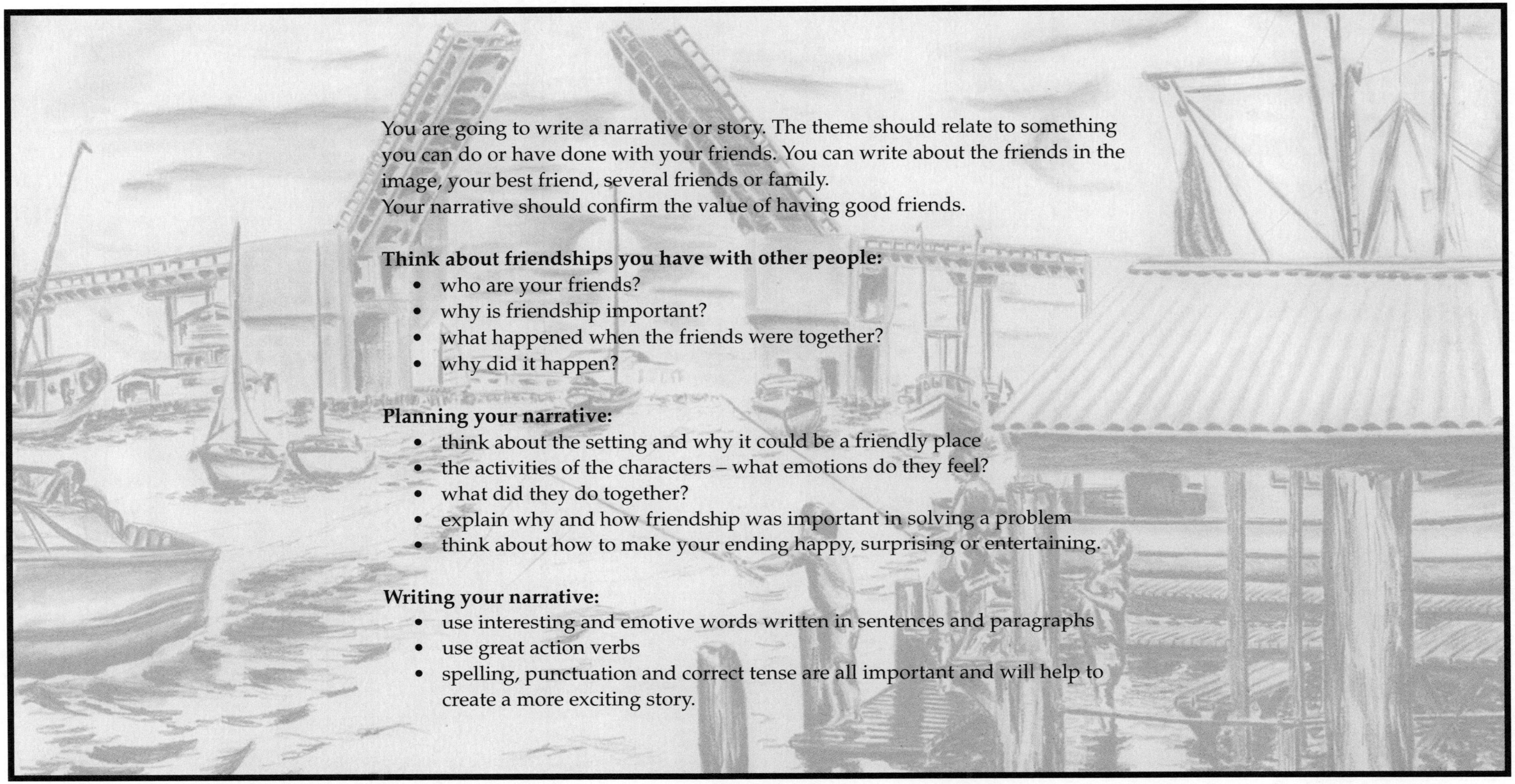

You are going to write a narrative or story. The theme should relate to something you can do or have done with your friends. You can write about the friends in the image, your best friend, several friends or family.
Your narrative should confirm the value of having good friends.

Think about friendships you have with other people:

- who are your friends?
- why is friendship important?
- what happened when the friends were together?
- why did it happen?

Planning your narrative:

- think about the setting and why it could be a friendly place
- the activities of the characters – what emotions do they feel?
- what did they do together?
- explain why and how friendship was important in solving a problem
- think about how to make your ending happy, surprising or entertaining.

Writing your narrative:

- use interesting and emotive words written in sentences and paragraphs
- use great action verbs
- spelling, punctuation and correct tense are all important and will help to create a more exciting story.

Writing Test – Narrative

Look at the stimulus. Brainstorm your ideas and then use this box to write a clear plan. Allow no more than five minutes. This planning will not be marked.

9780170462198

9780170462198

Glossary of Key Terms

Adjective
A word which modifies (adds more meaning to) a noun or a pronoun.
Adverb
A word which modifies (adds more meaning to) a verb, an adverb, or an adjective.
Alliteration
A sound device where the consonant sound at the beginning of a word is repeated a number of times in a sequence of words. For example, 'big, bright and beautiful'. In this instance, alliteration occurs using the letter 'b'.
Apostrophe
A punctuation symbol (') used to indicate a contraction or possession. For example, 'don't' is short for 'do not'. Another example is 'Alex's boots'. The apostrophe in 'Alex's' shows that the boots belong to Alex, indicating possession.
Climax
The pivotal point in a narrative, i.e. the most important or exciting event.
Cohesion
Describes the different language devices used to link words and sentences in a text smoothly and effectively to create meaning.
Comma
A punctuation symbol (,) used to indicate where a break might appear in a sentence. Commas may appear around a phrase or before a connecting word and the second part of a sentence. A comma can also be used to separate items in a list.
Complex sentence
Consists of one independent clause and one or more dependent clauses. Commas are used to separate sentence clauses.
Composer
The original creator of something, such as a text, musical piece, etc.
Compound sentence
Is composed of two simple sentences joined together by a comma and a joining word or coordinating conjunction. The seven coordinating conjunctions:

and	or	yet	nor
but	for	so	

Conjunction
Words like 'and', 'but', 'when', 'or', etc., which connect sentences, phrases or clauses.
Context
Background information that helps us to understand the text. Who is writing it, who is the intended audience and why are they writing it?
Controlling idea
The main argument or purpose for writing as stated in the introduction.
Convention
A rule, method, or practice established by usage or custom.
Deconstructing text
Involves the close reading of texts by breaking them down into components.
Dialogue
A conversation between two or more people.
Exposition (writing)
Argues a point of view.
First person
'I' and 'me'. Text written in the first person is personal and informal, and should therefore not be used in formal writing
Inference
A conclusion based on reasoning and evidence.
Informative (writing) Non-fiction report or descriptive text.
Language technique or device
Are used to expand meaning. For example, you could add adjectives to a noun, add adverbs to verbs. Phrases, similes, metaphors, etc. are also great devices that can be used effectively to illustrate a text.
Metaphor
A type of figurative language where the writer uses an implied comparison to add further meaning. For example consider the metaphor, 'Perfect storm for investment'. It is used to describe difficult financial circumstances rather than describe the weather, as the use of 'storm' adds potency to how difficult the circumstances are.
Narrative (writing)
The telling of a story.
Narrator
The person (named or unknown) who tells a story.
Personal pronoun
Is used to substitute the names of the people or things that perform actions.
Personification
A type of figurative language where an inanimate thing or object is given human qualities. For example, 'the tree's knobbly fingers reached out towards me'. In this example the tree's branches are described as human fingers.
Point of view
Describes the perspective or source of a piece of writing.
Pronoun
Is used in a sentence to replace a noun. They help the flow in writing by reducing the repetition of the same word. Pronouns come after the word they are replacing in a sentence. There are many different types of pronoun, including the personal, the demonstrative, the relative, the intensive, the reflexive, the interrogative, and the indefinite.
Pun
The use of a word, or of words which are formed or sounded alike but have different meanings, often for humorous effect.
Punctuation
Includes commas, full stops, quotation marks, etc. Punctuation helps to make the meaning of a text clear and therefore helps you to read a text.
Purpose
The reason for writing a text.
Recount (writing)
The retelling of an event.
Resolution
The part of a story that brings some conclusion to the complications in a narrative.
Rhetorical question
A question that does not require an answer because the answer is stated or obvious.
Scan
To quickly read or glance over.
Second person
'You'. The second person helps to create the sense that the writer is talking directly to you, so you feel engaged and involved in the text.

Sequencing
Positions events and things, into an order, which is logical and that you can easily follow
Simile
A type of figurative language where the writer compares two things using 'like', 'as' or 'than'.
Skim
To read, study, consider, etc., something in a surface or basic way.
Structure
The way a text is assembled or put together.
Symbol or motif
An object which represents something other than itself. Symbols or motifs may be used in writing as metaphors or repeated for cohesion.
Syntax
The way words are arranged in a sentence.
Tense
Is established through the verbs to specify whether an action happened in the past, present, future.
Thesis
The main argument or purpose for writing as stated in the introduction.
Third person
'He', 'she', 'it', and 'they'. The third person is more authoritative and objective than the first person ('I', 'me') or second person ('you').
Topic sentence
The first sentence of a paragraph that indicates what topic will be explored.
Viewpoint
The position of the narrator in relation to the story and the outlook of the events related in the story. There could be a first, second or third person narrator, for example.

9780170462198